LAKES: FORM AND FUNCTION

LAKES: FORM AND FUNCTION

Lars Håkanson

Department of Earth Sciences
Uppsala University
Sweden

Lakes: Form and Function

ISBN-10: 1-930665-24-5
ISBN-13: 978-1-930665-24-8

Library of Congress Control Number: 2004109155

THE BLACKBURN PRESS
P. O. Box 287
Caldwell, New Jersey 07006 U.S.A.
973-228-7077
www.BlackburnPress.com

To Lena, Kalle, Nisse and Olle

with love

Contents

LAKES: FORM AND FUNCTION

1. Background, Introduction and Aim

The first classical work with a specific focus on the study of lake forms, describing their genesis and role in a broad limnological perspective, was "Limnological Methods" by Welch (1948). The source of morphometric data quantifying the size and form of lakes is the bathymetric map and such maps are today generally derived from echosounding, a technique that became more widely used after 1948. It is not mentioned in Welch's book. My book from 1981, "A Manual of Lake Morphometry" (Håkanson, 1981a), is essentially a "cookbook" on how to conduct echosoundings in lakes, to evaluate the echograms, to construct bathymetric maps and to define morphometric parameters. Examples are given why this is important in lake studies, but those examples are not the focal point of that book. Most modern textbooks in limnology (see, e.g. Horne and Goldman, 1994; Wetzel, 2001 and Kalff, 2002) treat lake morphometry as if this were essentially a descriptive subject; and relatively few attempts have been made to demonstrate why morphometry is of fundamental importance for the understanding of lake ecosystem functioning, i.e., for the structure and function of lake foodwebs and in regulating fluxes of substances in lakes. The aim of this book is to do just that!

This book will also discuss Geographical Information Systems (GIS) and how GIS can be used to calculate important lake morphometric parameters in a quick and objective manner. GIS was not mentioned in my 1981 book.

The way in which morphometric parameters are defined and abbreviated in modern textbooks can illustrate the confusion and perspectives characterizing prevailing viewpoints in limnology on this subject. Table 1.1 demonstrates this point clearly:

Table 1.1. A comparison of different symbols and abbreviations used in three standard textbooks in limnology, Wetzel (2001), Horne and Goldman (1994) and Kalff (2002); and the system advocated and used in this book (see also table 1.2).

	Wetzel	Horne& Goldman	Kalff	This book
Maximum length	l	-	L	L_{max}
Maximum breadth	b	-	-	B_{max}
Area	A	A	A	A
Volume	V	V	V	V
Maximum depth	z_m	z_{max}	Z_{max}	D_{max}
Mean depth	$\bar{z}$	$\bar{z}$	$\bar{z}$	D_{mv}
Relative depth	z_r	-	z_r	D_{rel}
Shoreline length	L	L	L'	L_0
Shore development	D_L	D_L	D	L_d
Volume development	-	-	-	V_d
Mean slope	-	-	S	S_{mv}

Capitals and small letters are used more or less arbitrarily, as are the subscripts. This usage may not be a crucial matter if lake morphometric data are rarely used, or only used for simple descriptions; but if such data are used in contexts of mass-balances or to understand why different lakes respond differently to nutrients or contaminants and why the foodweb characteristics differ among lakes, then there is a strong need to use and abbreviate these morphometric parameters in a logical and rational way. Table 1.2 gives a compilation of abbreviations used in this work.

Table 1.2. A general system for abbreviations of lake morphometric parameters and variables related to dynamic mass-balance models.

A. Overall criteria:

- No Greek letters accepted.
- Abbreviations should be logical, consistent, easy to understand and as self-explanatory as possible.

B. Lake morphometric parameters:

- Length measures (L): maximum length = L_{max}; maximum effective length = L_{xel}; effective length = L_{el}; effective fetch = L_{ef}; shoreline length = L_O; shoreline length of islands = L_{is}; total shoreline length = L_{tot}; contour line length at 5 m depth = L_5.
- Breadth measures (B): maximum breadth (or width) = B_{max}; maximum effective breadth = B_{xeb}; mean breadth = B_{mv}.
- Depth measures (D): maximum depth = D_{max}; mean depth = D_{mv}; median depth = D_{50}; quartile depths = D_{25} and D_{75}; relative depth = D_{rel}; depth of the wave base = D_{wb}.
- Area measures (A): total lake area = A_{tot}; water surface area = lake area = A; area below 5 m depth = A_{5b}; area above 5 m depth = A_{5a}; 3D-bottom area = A_{3D}.
- Volume measures (V); water volume = V; volume below 5 m depth = V_{5b}; volume above 5 m depth = V_{5a}.
- Slopes (S): the slope between two contour lines (5 and 10 m) = $S_{5/10}$; mean slope = S_{mv}; median slope = S_{50}.
- Shore development = shoreline irregularity = L_d.
- Lake bottom roughness = Ro; form roughness = Ro_f.
- Volume development = form factor = V_d.
- Insulosity = Ins.
- Dynamic ratio = DR.
- Filter factor = FF.

C. Abbreviations related to dynamic mass-balance models:

- C for concentrations (g/l = M/V, mass per volume, etc.), F = fluxes (g/time), mass = M (g), rates = R (1/time), velocities = v (R = v/D; the rate is equal to a velocity divided by a depth; v in m/time), distribution coefficients = DC.
- Fluxes from compartment A to compartment B = F_{AB}; concentration in A = C_A; mass in A = M_A; rate regulating flux from A to B = R_{AB}; distribution coefficient regulating flux to either A or B = DC_{AB}.

There are a few simple rules:

- No Greek letters have been accepted in spite of the fact that such nonsensical symbols are frequently used, in particular in physical limnology.
- All abbreviations should be self-explanatory, clear and logical: Length measures are abbreviated as L with an index for the given measure, for example, L_{max} for the maximum length, B for breadth measures, D for depth measures, V for volumes, A for areas, S for slopes; dimensionless measures are abbreviated so that they are understandable, for example, DR for the dynamic ratio and V_d for the volume development.
- The same arguments may be given for abbreviations in mass-balance calculations: C for concentrations, F for fluxes, M for mass, R for rates and DC for distribution coefficients. So, F_{AB} means a flux from compartment A to compartment B.

This volume is not a textbook in modelling; and there are, for example, no differential equations in the running text. It is enough to know addition, subtraction, multiplication and division to follow this text. But it is not possible to understand how lakes function without a basic knowledge of the transport processes that exist in all lakes and apply to all substances. Fig. 1.1 gives a compilation of such processes; and it will be shown that lake morphometric data are fundamental for all these processes, except tributary inflow!

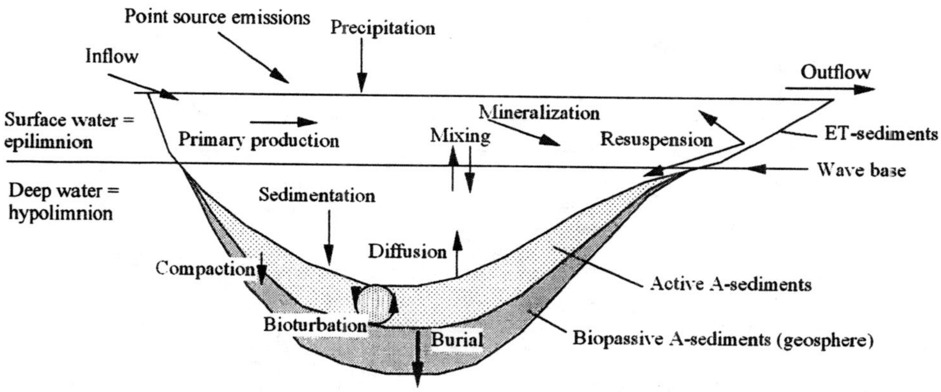

Fig. 1.1. Illustration of general and fundamental transport processes to, within and from lakes.

These processes have the same names for all lakes and for all substances; so

- sedimentation is the flux from water to sediments,
- resuspension is the advective flux from sediments back to water,
- diffusion is the flux from sediments back to water triggered by concentrations gradients,
- mixing appears in stratified lakes and is the transport between deep-water areas and surface-water areas,
- mineralization is decomposition of organic particles by bacteria,
- primary production is creation of living suspended particles from dead matter by sunlight,
- burial is the transport of matter from the lake biosphere to the geosphere often of matter from the technosphere,
- outflow is the flux out of the lake of water of everything dissolved and suspended in the water.

The size and form of a lake influence all these processes, which will be demonstrated by many examples in this book.

Fig. 1.2 exemplifies that lake morphometric characteristics are crucial for benthic organisms; zoobenthos, benthic algae and macrophytes are functional groups of paramount importance in most lakes.

The production and biomasses of these organisms depend on lake form and water clarity, which, in turn, depends on lake form. This will be demonstrated by many examples in chapter 5. Shallow lake areas may be regarded as a "pantry and a nursery" for the entire lake. It has been demonstrated (see fig. 1.3) that shallow coastal areas in the Baltic can have a very high bioproduction capacity (20 times higher than the most productive areas on land; see Rosenberg, 1985); all three functional groups of primary producers - phytoplankton, benthic algae and macrophytes - are present in such areas (but not in open water areas); and where there is a high primary production, there is also a high secondary production of zooplankton, zoobenthos and fish (see Håkanson and Boulion, 2002).

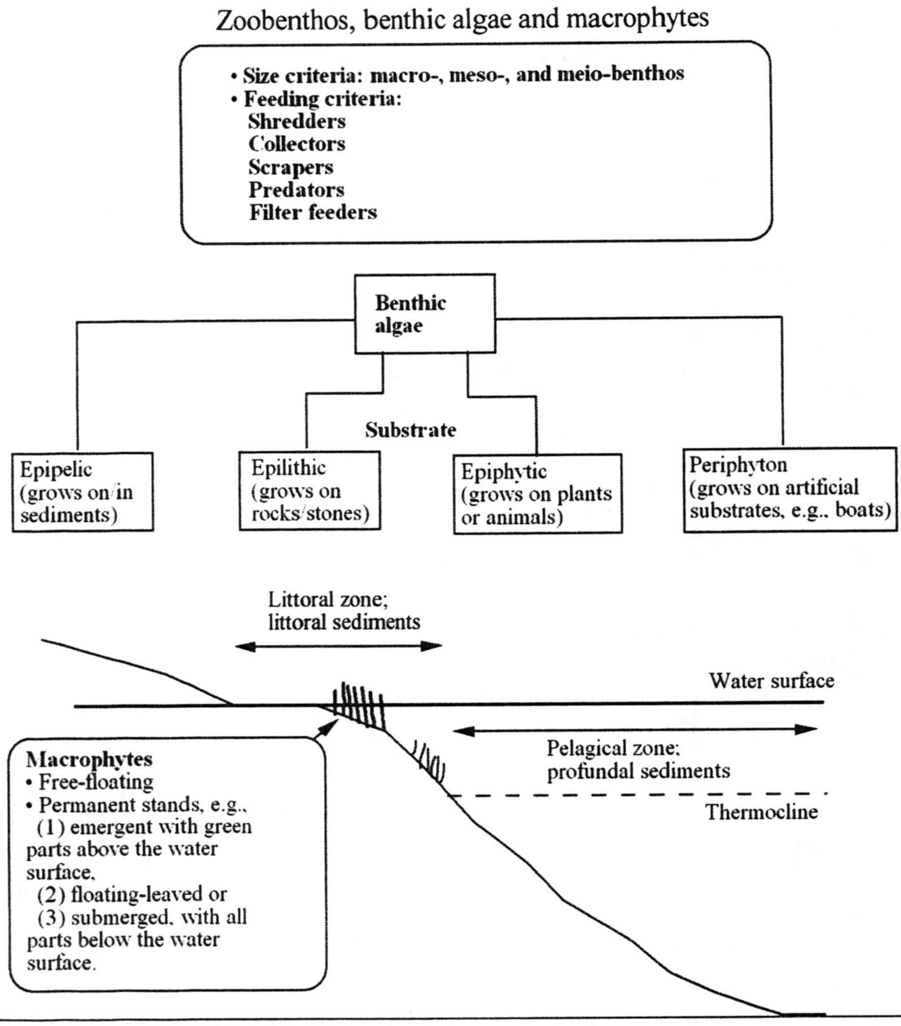

Fig. 1.2. Compilation of concepts related to lake morphometry and sediment-living organisms (zoobenthos, benthic algae and macrophytes; see also Vollenweider, 1968, 1976; Cummings, 1973; Brinkhurst, 1974; Wetzel, 2001).

From fig. 1.3, one can note that at water depths larger than 15 m, one can generally set the production capacity (PR/BM, PR = production in kg/time and BM = biomass in kg) of the infauna (i.e., animals > 1 mm living in the sediments) to be about 1. At water depths of 3 - 15 m, PR/BM is generally between 1 and 3. At water depths smaller than 3 m, it is important also to consider the sediment type, the habitat for the infauna. PR/BM is given on a relative scale of value in terms of fishery biology:

- • means a valuable coastal area,
- •• means a very valuable area and
- ••• an extremely valuable area.

Lake morphometry does not, however, influence "everything" in lakes. So, for which lake variables can one expect a great influence from lake size and form characteristics? This subject will be discussed using empirical data in chapter 4. Fig. 1.4 is included here to stress one point:

To analyze the relative importance of lake morphometry in relation to other factors that may influence fundamental lake variables, such as properties of the catchment area, climatological factors, etc., it is important to realize that any lake variable (y) will depend on factors causing seasonal variations within lakes and on factors causing variations in mean, median or characteristics values among lakes. Lake morphometry will generally influence variations among lakes; but also many internal processes (see fig. 1.1). This means that a given load to a lake, for example, of nutrients and/or toxins, may cause very different effects in lakes of different morphometry. Effect-load-sensitivity analyses are of fundamental importance in lake management, and this will subject be addressed in chapters 4 and 5.

So, the same load of a contaminant may cause different effects in different lakes. Then, one can ask: How much of such variations among lakes may be statistically related to morphometric parameters (see fig. 1.4C)? This topic will be discussed in chapter 4 using empirical data for many standard lake variables and data from many lakes.

Chapter 2 will focus on echosounding and bathymetric map construction. All morphometric data, whether calculated using manual methods or GIS, emanate from a bathymetric map. There is a close relationship between the way in which the survey is carried out (the intensity and accuracy of the survey), the error in the bathymetric map and the information value of the map. If there are no contour lines in the map, the information on the topographical features of the given lake is zero (see fig. 1.5).

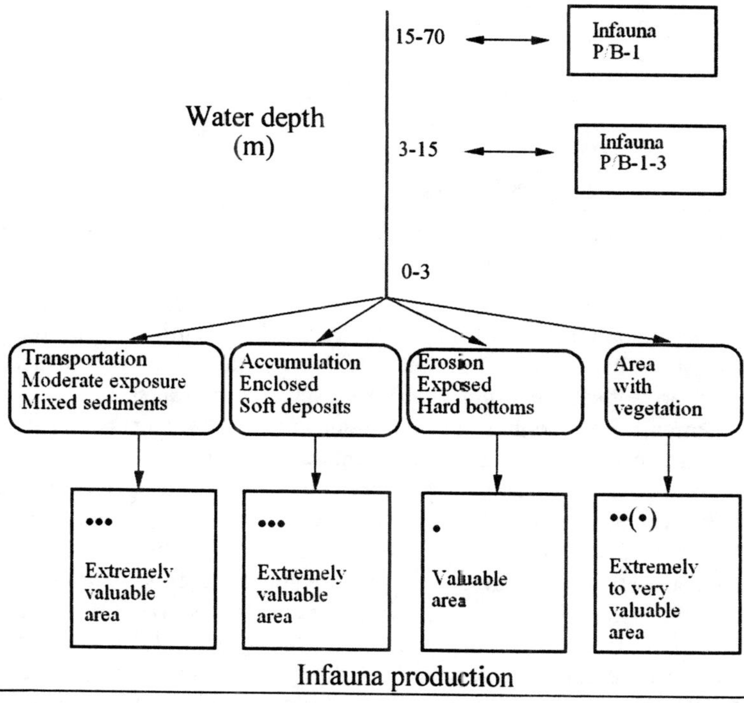

Fig. 1.3. Criteria to estimate the "value" of Baltic coastal areas mainly from a perspective of fishery biology.

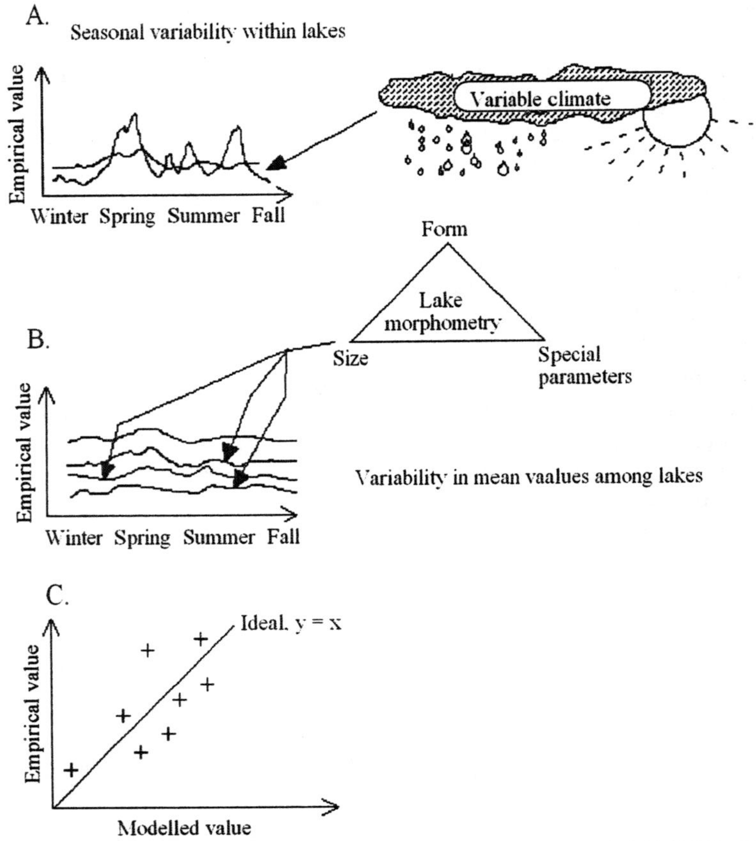

Fig. 1.4. Illustration of factors influencing within- and among-site variability in a lake variable and the role of lake morphometry for variations among lakes and climatological variables for variations within lakes.

The information value (I) increases rapidly with the first number of contour lines; but if the number of contour lines becomes high enough, the gain in information becomes negligible, even if more contour lines are added. This subject will be discussed in chapter 2.

Chapter 3 presents and defines a great number of lake morphometric parameters and exemplifies why and how these parameters are important in lake studies.

Lakes may also be classified and categorized in many ways using different criteria. From a geological perspective, lakes may be looked upon as causal objects on the face of the Earth. As all geomorphological features, lakes undergo different stages: Creation, young, maturity, old age, extinction. Lakes are generally created by drastic geological events, such as volcanism, tectonic activity or glaciation. The relief of the basin is under constant alteration by exogenic forces, such as weathering, erosion, transport and deposition. Practically all 227,000 lakes in Sweden (objects larger than 1 ha; see table 1.3) have been influenced by the enormous form-creating power of the last (but not latest) glaciation. Eventually, all these lakes will be filled up and vanish; and the evident relationship between lake size and number of lakes in each size class will disappear.

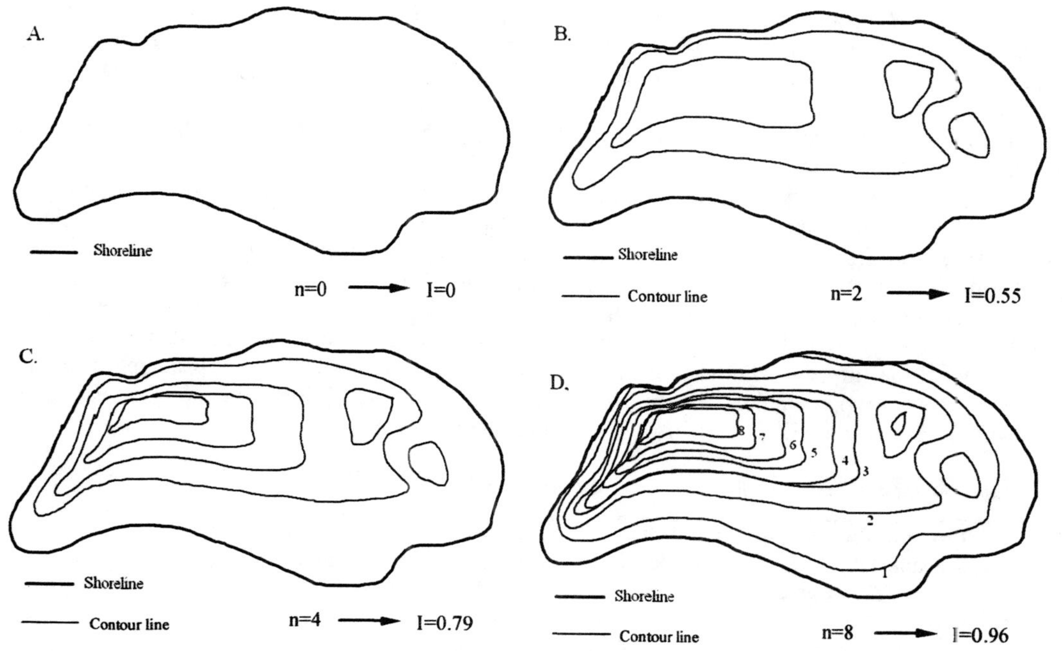

Fig. 1.5. Illustration of how the information value (I) increases with the number of contour lines (n) in a bathymetric map.

Table 1.3. The number of lakes in Sweden arranged by size (data from Monitor, 1986 and Håkanson and Peters, 1995).

Area ($1\ km^2 = 100$ ha)	**Number**	**Cumulative number**
>1000	3	3
100 - 1000	19	22
10 - 100	362	384
1-10	3987	4371
0.1 - 1	19374	23745
0.01 – 0.1	59500	83245
0.001 – 0.01	144000	227000

Hutchinson (1957) has, in a brilliant chapter, discussed all major lakes types, their origin, morphometric characteristics and distribution on Earth. He differentiated between eleven major lake types that were divided into 76 sub-types. The eleven major lake types are summarized in table 1.4.

This book focuses on form and function, and table 1.5 gives a classification of lake types based on trophic level.

Ultraoligotrophic (very low-productive) lakes are characterized by mean summer chlorophyll-a concentrations lower than 0.1 µg/l and by fish such as char and trout. Oligotrophic (low-productive) lakes have chlorophyll-a concentrations in the range from 0.1 to 1 µg/l and fish like trout and whitefish. One can note the characteristic level of total phosphorus overlap between the classes (see also OECD, 1982). The fish yield in mesotrophic lakes is about 0.9 g wet weight (ww) per m^2 per year, and by multiplication with lake area and 365 one can calculate an order-of-magnitude value on the annual fish production using this information. Eutrophic (high-productive) and hypertrophic (very high-productive) lakes have higher primary and secondary production. Chapter 5 will discuss how lake bioproduction is related to lake morphometry.

Table 1.4. Major lake types on Earth according to Hutchinson (1957).

1. Tectonic lakes, e.g., basins in faults (like Lake Bajkal and Lake Tanganyika).
2. Volcanic lakes, e.g., maars, caldera lakes and lakes formed by damming of lava flows.
3. Landslide lakes, e.g., lakes held by rockslides, mudflows and screes.
4. Glacial lakes:
   A. Lakes in direct contact with ice, e.g., lakes on or in ice and lakes dammed by ice.
   B. Glacial rock basins, e.g., cirque lakes and fjord lakes.
   C. Morainic and outwash lakes, e.g., lakes created by terminal, recessional or lateral moraines.
   D. Drift basins, e.g., kettle lakes and thermokarst lakes.
5. Solution lakes, e.g., lakes formed in caves by solution.
6. Fluvial lakes:
   A. Plunge-pool lakes.
   B. Fluviatile dams, e.g., strath lakes, lateral lakes, delta lakes and meres.
   C. Meander lakes, e.g., oxbow lakes and cresentic levee lakes.
7. Aeolian lakes, e.g., basins dammed by wind-blown sand and deflation basins.
8. Shoreline lakes, e.g., tombolo lakes and spit lakes.
9. Organic lakes, e.g., phytogenic dams and coral lakes.
10. Anthropogenic lakes, e.g., dams and excavations made by man.
11. Meteorite lakes, e.g., meteorite craters.

Table 1.5. Trophic classification of lakes based on fixed steps (a factor of 10) of mean summer chlorophyll-a concentrations (Chl), typical ranges for total phosphorus concentrations (TP), phytoplankton primary production (PrimP in µg C/l·d and g ww/m^2·yr; a factor of 2 is used to calculate g C to g dw; 5 to calculate g dw to g ww; the effective depth of the photic zone is equal to the Secchi depth; and the duration of the growing season is set to 150 days); and dominating species of fish in the given classes. Also note that primary and secondary production in highly eutrophic lakes will be reduced by self-shading of algal cells (from Håkanson and Boulion, 2002).

Trophic level	**Chl (µg/l)**	**Range TP (µg/l)**	**PrimP (µg C/l·d)**	**Characteristic PrimP range (g ww/m²·yr)**	**Characteristic Fish yield (g ww/m²·yr)**	**Dominating fish Species**
Ultraoligotrophic	≤0.1	<3	≤4.4	≤125	0.08	Char, trout
Oligotrophic	≤1	1-10	≤32	125-380	0.3	Trout, Whitefish
Mesotrophic	≤10	5-40	≤270	380-1400	0.9	Whitefish, Perch
Eutrophic	≤100	20-400	≤2400	1400-6400	3	Perch, Roach
Hypertrophic	≤1000	>100	>2400	>6400	7	Roach, Bream

2. Methodological Aspects

This chapter draws much from a similar chapter in Håkanson (1981a). The aim here, however, is not to repeat that information but rather to highlight key information regarding echosounding and bathymetric map construction and to underscore the information value of the bathymetric map. This chapter will also give a short background on GIS.

Hydrographic surveys can be carried out for many purposes, for example, to meet requirements for safe navigation. In such cases every detail in the map has to be correct. For fishing, it may be of interest to get information about the bottom roughness and obstacles that may damage the fishing equipment. In lake studies, it is often a primary interest to have a bathymetric map illustrating the general morphometric characteristics of the lake, as well as topographical features influencing the representativity of water or sediment samples.

2.1. Echosounding

Many types of commercially available instruments are available both for fixed installations on larger boats (including multi-beam emission instruments) and more simple instruments for mobile use in small boats in lakes. The principle of an echosounder is shown in fig. 2.1. Basically, the equipment consists of (1) a transmitter, (2) a sender-receiver unit for the acoustic wave, (3) an amplifier unit, (4) a recording unit and (5) a mechanical suspension unit. Most instruments are fed by 12 volts DC power from an accumulator. They generally work with frequencies in the range from 10 to 250 kHz. The low frequencies (< 50 k Hz) provide better sediment penetration and hence also better information on the physical sediment characteristics than do instruments using high frequencies.

The crystal in the underwater sender-receiver unit is fed by the pulses from the transmitter and forced into vibrations. These vibrations are propagated in the water as acoustic waves, which are sent out in the form of a cone and not as straight lines. The base of the cone becomes wider with increasing water depth and the reflected signal gets weaker with increasing depth. The reflected signal is caught by the receiver, then amplified and registered on the echogram as a depth value. The acoustic wave has a velocity of about 1.5 km/s in salt water. The equipment measures the time it takes for the sonic wave to go down and up. Since the velocity is practically constant, the recording unit is graded in length units and not in time units. Some instruments can be calibrated for water temperature and salinity by means of a frequency meter affecting the velocity of the recording pen.

The instrument should be placed in the middle of the boat and at about 1.5 m water depth to prevent it from coming above the water surface during the survey. The zero line of the echogram should be adjusted to the actual depth value. The underwater unit should not be placed close to a keel where the acoustic wave may be shadowed, and not where air bubbles may affect the signal. It is of utmost importance to control the sender-receiver unit so that it is unaffected by vegetation that may reduce the functional ability.

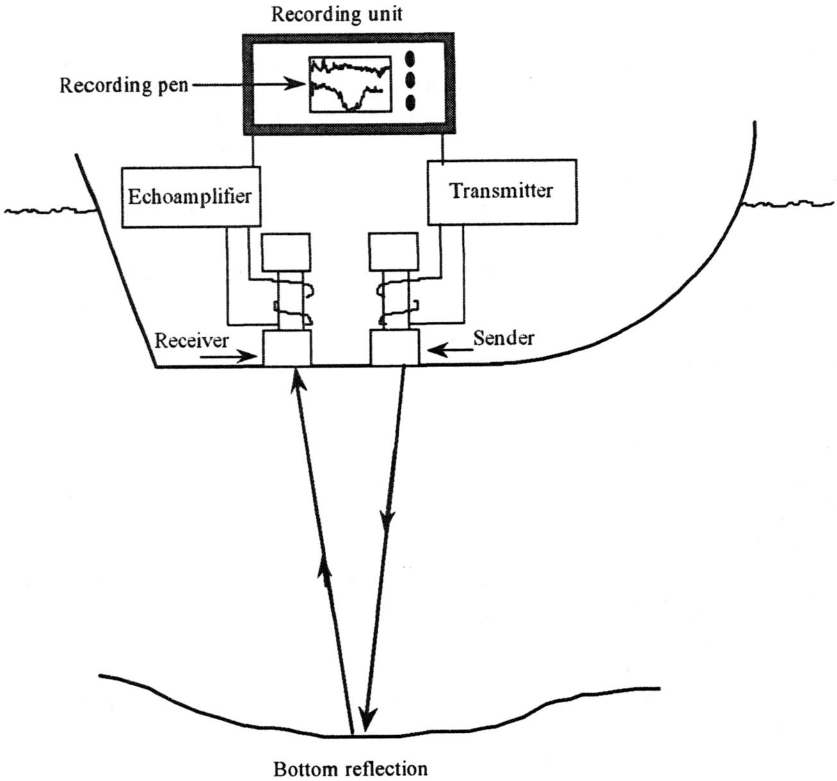

Fig. 2.1. Illustration of functional principles of an echosounder.

2.1.1. Fieldwork

Complementary field equipment includes a graded sounding line with a lead weight or a sounding rod for use in shallow waters. Today, one can determine the position very accurately by GPS. One also needs an accurate map in a relevant scale. To optimize the information from the survey, it is recommended that a system of crossing tracks is used. The intensity of the survey, i.e., the average distance between the tracks (the L_r value, see later) should be matched to the size of the area to be surveyed and its topographical roughness. It is also beneficial to conduct a pilot survey to identify possible obstacles. It is important to have the instrument duly calibrated, for example, against data from a sounding line. A technique with "flying start" should be practiced, and two persons are generally needed for the survey. During the measurement, the boat should keep a constant speed and course. In the survey journal, one should mark where the tracks start and end, as well as deviations from the planned course.

2.1.2. Map construction

After the survey, the echogram for each track is cut into a separate strip, and the depth intervals that are to be used on the requested bathymetric map should be determined. The basic principle is that the contour line interval (L_{cl} in m) should be positive integers and the relationship between L_{cl} and the number of contour lines (n) may be given by the ratio between the maximum depth (D_{max}) and n, i.e., $L_{cl} = D_{max}/n$. For most limnological studies, as when bathymetric maps for whole lakes are requested, one can use between 4 and 20 contour lines. The next step is to mark all points corresponding to the selected depth values (see fig. 2.2). From these points parallel lines are drawn which cross the baseline (the zero line) at right angles. This information is then transferred to the map.

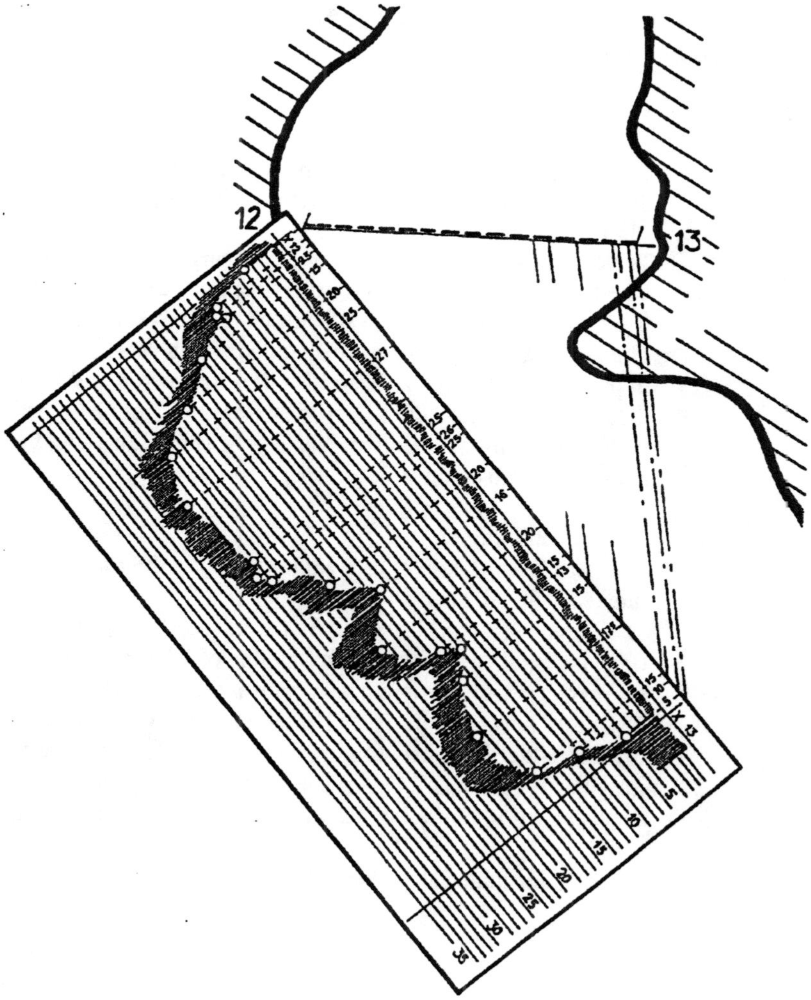

Fig. 2.2. Example of a transformation of depths from an echogram to a bathymetric map.

2.2. The information value of bathymetric maps

Bathymetric maps are the source of the morphometric parameters. The reliability of these parameters depends on the accuracy of the bathymetric map, which in turn depends on the intensity and accuracy of the survey. The aim of this section is to discuss the relationship between the intensity of the survey and the yield of the survey (for further information on this topic, see Håkanson, 1978a).

The following relationships are included in the optimization model, which is based on echograms along given tracks.

$$I = I' \cdot I'' \quad (2.1)$$

$$I' = [A - 0.14 \cdot L_r \cdot L_d^2 \cdot (1/(n+2))^{0.5} \cdot \Sigma \sqrt{A_i}]/A \quad (2.2)$$

$$I'' = (e^{0.4 \cdot n} - 1)/(e^{0.4 \cdot n} + 0.02) \quad (2.3)$$

Where

I = The information value of the bathymetric map; a figure between 0 and 1; if the information in the bathymetric map is complete and completely correct, I is 1.

I' = The correctly identified area in the bathymetric map. I' also varies between 0 and 1; it is 1 when all contour lines are correct. Then, the area error (ER) is zero. These concepts are illustrated in fig. 2.3. So, I' = I – ER.

I'' = The information number which depends on the number of contour lines (n) in the bathymetric map. Also I'' varies between 0 and 1. Table 2.1 gives data calculated from eq. 2.3. I'' describes the fact that a bathymetric map with fewer contour lines provides less information about the lake than a map with more contour lines.

A = The lake area (km^2).

A_i = The total (= cumulative) area (in km^2) within the limits of a given contour line (see table 2.2).

L_r = The distance (in km) between the tracks (see fig. 2.4 for information). For large lakes ($A > 1000$ km^2), an L_r value of 2 km would provide good input for map construction, but a very poor input for small lakes (< 1 km^2).

L_d = The shore development (see chapter 3). This dimensionless parameter expresses the shore irregularity, and it is highly corrected to the bottom irregularity (see fig. 3.43). A lake with irregular topography should be surveyed with greater intensity to obtain a map with the same information value.

e = The base for natural logarithms; e = 2.718.

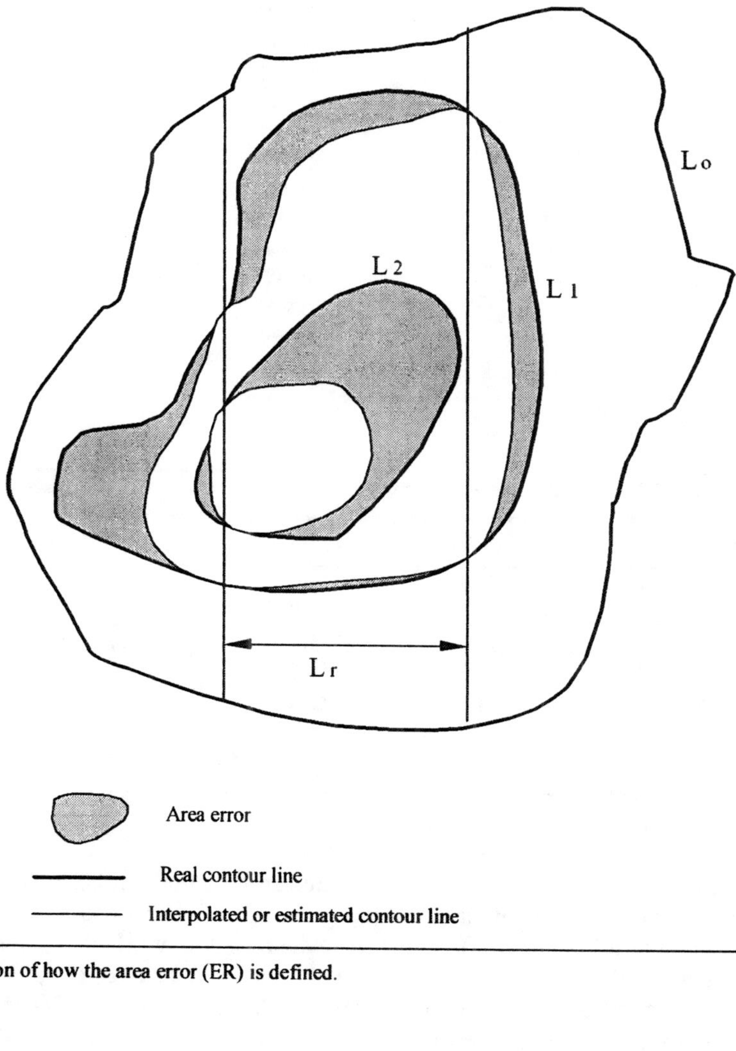

Fig. 2.3. Illustration of how the area error (ER) is defined.

Table 2.1 The relationship between the number of contour lines (n) and the information number (I").

n	I"
1	0.325
2	0.546
3	0.695
4	0.795
5	0.862
6	0.908
7	0.938
8	0.959
9	0.971
10	0.981
12	0.992
14	0.996
16	0.998
18	0.999
20	0.9997
30	1.000

Table 2.2. Cumulative areas for a hypothetical lake.

n	A_i (%)	A_i (km^2)	$\sqrt{A_i}$ (km)
1	75.2	375	19.4
2	51.6	258	16.1
3	30.7	153.5	12.4
4	14.3	71.5	8.5
5	4.4	22.0	4.7
		Sum:	61.0

The proportion of correctly identified area (I') increases as the number of contour lines (n) decreases, but as the number of contour lines decreases, there is a reduction in the information value (I''). The optimal information (I) is obtained when the product I'·I'' attains its maximum value. The information value (I) is related to (1) the intensity of the survey (L_r), (2) the size of the lake (A), (3) the topographical irregularity of the basin (L_d), (4) the number of contour lines in the bathymetric map (n) and (5) the lake form (as given by $\Sigma\sqrt{A_i}$). The sequence just given is also the order in which these factors influence I.

2.2.1. The intensity of the survey

If the survey is conducted along parallel tracks separated by a distance L_r, then L_r may be used directly as a measure of the intensity of the survey. If the soundings are done from a regular square net, a representative L_r value may be determined from the ratio between the lake area (A) and the total length of the echosounding tracks, i.e., $L_r = A/\Sigma L_i$. If the survey is done according to an irregular net, where the tracks are not parallel or crossing at right angles, a representative L_r value may also be calculated from the given ratio provided that the tracks give an even area coverage of the whole lake surface.

If, finally, the lake has been surveyed in such a manner that certain parts of the lake have been measured more intensely than other parts, fig. 2.4 illustrates how to calculate a representative L_r value. Assume that A = 100 km^2 and the total length of the echosounded tracks is 62.4 km. Then, the uncorrected L_r value is 1.60 km, the corrected value is 1.86 km. The longest of the crossing tracks are used as references and these tracks are fully counted. All tracks crossing these are reduced by a factor $\sin(a_i)$, where a_i is the acute angle between the tracks.

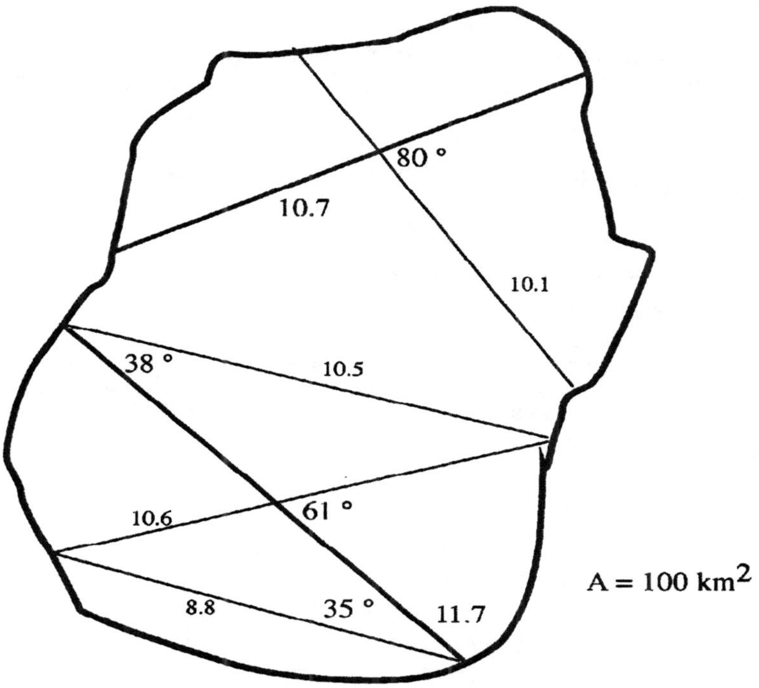

	Length (Li)	Angle (ai)	sin(ai)	Li·sin(ai)
	10.7	-	1	10.7
	10.1	80	0.985	10.5
	11.7	-	1	11.7
	10.5	38	0.616	6.5
	10.6	61	0.875	9.3
	8.8	35	0.573	5.0
Sum:	62.4			53.7
A/Sum:	1.60			1.86 = Lr

Fig. 2.4. Example showing a method to determine representative L_r-values.

2.2.2. Practical use of the optimization method

This section will exemplify a typical and practical situation where this optimization approach can be applied. Empirical data will be used from Lake Vänern, Sweden. The lake is a reference lake in chapter 3. The lake area is 5648 km^2 (see table 2.3) and the average distance between the echosounded tracks was 500 m. The result of the survey is a bathymetric map in scale 1:180,000 with 10 contour lines. The following questions concerning this bathymetric map can now be discussed:

- What is the area error (ER) in the bathymetric map in scale 1:180,000 with 10 contour lines, and how much of the area is correctly identified (I')?
- What is the optimal information value (I) from a survey with $L_r = 0.5$ km?
- What is the optimal number of contour lines in the bathymetric map?

The area error (ER) is first calculated from I' = (1 - ER) using the following lake data: A = 5648 km^2, $L_r = 0.5$ km, $L_d = 7.14$ (see chapter 3), n = 10 and $\Sigma(\sqrt{A_i}) = 284.65$ (see table 2.3). Then:

$$I' = [5648-0.14\cdot 0.5\cdot 7.14^2\cdot(1/(10+2))^{0.5}\cdot 284.65]/5648 = 0.948$$

That is, 94.8% of the lake area is correctly identified and the area error is 5.2% or 294 km^2.

Table 2.3. Morphometric data on area and depth for Lake Vänern, Sweden.

Depth (m)	Cumulative area (A_i; km^2)	$\sqrt{A_i}$ (km)	n
0-10	5648	75.15	0
10-20	3995	63.21	1
20-30	2975	54.38	2
30-40	2234	47.26	3
40-50	1604	40.05	4
50-60	1026	32.03	5
60-70	544.9	23.34	6
70-80	209.0	14.46	7
80-90	48.94	7.00	8
90-100	5.36	2.32	9
100-106	0.37	0.61	10
	Sum:	284.65	

The total length of the contour lines is 11,400 km. This means that the average error in length units perpendicular to the contour lines is 294/11400 = 0.026 km, or in scale 1:180,000, 0.14 mm, which is about equal to the width of the contour lines on the map. Thus, the error may, for most practical purposes, be regarded as negligible.

The optimal information value (I) is determined in table 2.4, which shows that the highest I value would be for a map with 13 contour lines. That I value is 0.935, which is close to the I value for the map with 10 contour lines (0.930). A map with 20 contour lines would give a higher information value (I'') but a comparatively high area error (ER). Real topographical features vanish if too few contour lines are being used, and irrelevant features may appear if too many contour lines are used.

Table 2.4. Optimal information value (I) and optimal number of contour lines (n) related to the bathymetric map for Lake Vänern.

Contour line interval (m)	n	I"	$\Sigma\sqrt{A_i})$	I'	I
50	2	0.546	32.64	0.990	0.541
30	3	0.696	72.92	0.979	0.682
25	4	0.795	93.97	0.976	0.776
20	5	0.862	125.4	0.970	0.836
15	7	0.938	178.3	0.949	0.890
12	8	0.958	232.6	0.954	0.913
10	10	0.981	284.7	0.948	0.930
9	11	0.987	320.6	0.944	0.932
8	13	0.994	366.4	0.940	0.935
6	17	0.999	499.6	0.928	0.927
5	21	1.000	606.4	0.920	0.920
4	26	1.000	768.0	0.908	0.908

2.3. GIS applications

Today, there are many modern digitized methods applicable in lake studies including Geographical Information Systems (GIS). Those methods enable quick and effective estimations of lake morphometric variables (see Pilesjö et al., 1991), but many traditional, manual methods are also useful (see Wetzel, 2001). In contexts of GIS applications, there are also many commercial software programs available, including Surfer, ArcInfo, MapInfo and ArcView.

Since the morphometric parameters are measurements of different types (depths, volumes, slopes and lengths), it is not obvious which type of GIS data to use in the calculations. Vector data are often preferable when calculating lengths and areas, while raster data are preferable when calculating, for example, volumes and slopes.

The first step in the GIS calculations is to transform and store the depth data on the charts to digital vector and raster data. Since the map projection of most charts is not correct in length and area (conic projection), all coordinates should be transformed to a Universal Transversal Mercator (UTM) projection. The irregular depth (elevation) points from the vector file may then be interpolated to a gridded digital elevation model (raster data). Burrough (1986) and Skidmore (1990) discuss methods for interpolating irregular elevation points to a regular grid. Kriging algorithms are generally used for interpolations.

Special considerations should be given to coastal areas in large lakes. After digitizing and interpolation, the area outside the coastal area can be masked, resulting in one vector file containing coordinates for all shore lengths (of islands and the coastline) and openings to the surrounding open water areas, and one raster file containing the pixel size (header) and the digital elevation model of the coastal area. Fig. 2.5 illustrates steps to define a coastal area and the boundaries to the open water areas or adjacent coastal areas at the topographical bottlenecks using GIS techniques where the exposure ($Ex = A_{sa}/A$) attains minimum values when different alternatives for settling the boundary lines are tested (A_{sa} = section area; A = enclosed coastal area; method from Pilesjö et al., 1991). Once the coastal area is defined, one can also define important variables for mass-balance calculations, such as the coastal volume (regulating the concentration of any given substance), important morphometric parameters for internal fluxes, such as the mean depth and the water surface area, and key variables regulating the water exchange between the coast and the sea, such as the section area and the exposure (see fig. 2.5).

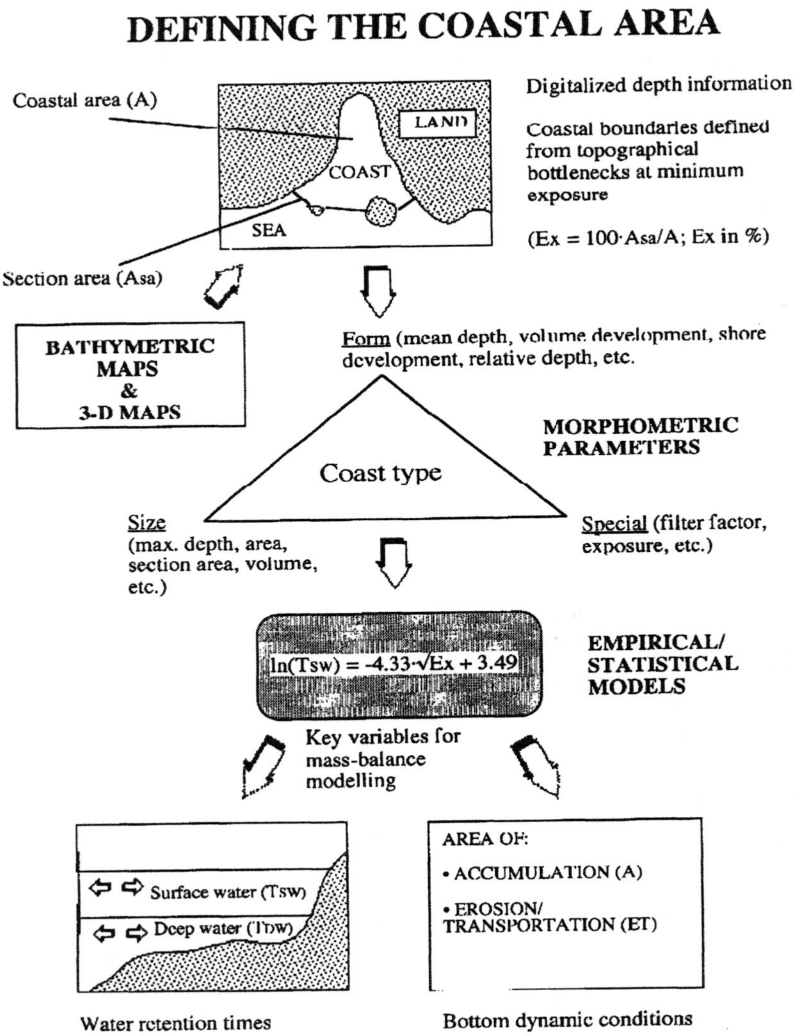

Fig. 2.5. Illustration of steps using GIS methods to define a coastal area in a large lake using the topographical bottleneck approach, as well as how parameters expressing lake or coastal morphometry may be calculated from the GIS map and then further processed into key variables for mass-balance calculations using empirical regressions to express surface and deep water retention times and bottom dynamic conditions.

This method of defining coastal areas also opens a possibility to use empirical models to estimate, for example, the theoretical water retention times of the surface water (see fig. 2.5) and the deep water, and the bottom dynamic conditions (regulating sedimentation, resuspension and diffusion) from morphometrical parameters describing the coastal areas (such as area, mean depth and section area). From the theoretical surface water retention time (T_{SW} in months), one can calculate the surface water flux (Q_{SW} in m^3/month) as V_{SW}/T_{SW}, where V_{SW} is the surface water volume (m^3) and the flux of the given substance, $Q_{SW} \cdot C_{sea}$ in g/month, where C_{sea} is the concentration of the substance in the sea outside the given coastal area (g/m^3). Fig. 2.6 shows the two main coast types in large lakes, (i) open coasts and (ii) archipelago coasts and bays, and how the theoretical surface water retention time (T_{SW}) may be calculated for each type and for any given coast which may be a combination of the two (for further details on these matters, see Håkanson, 2000).

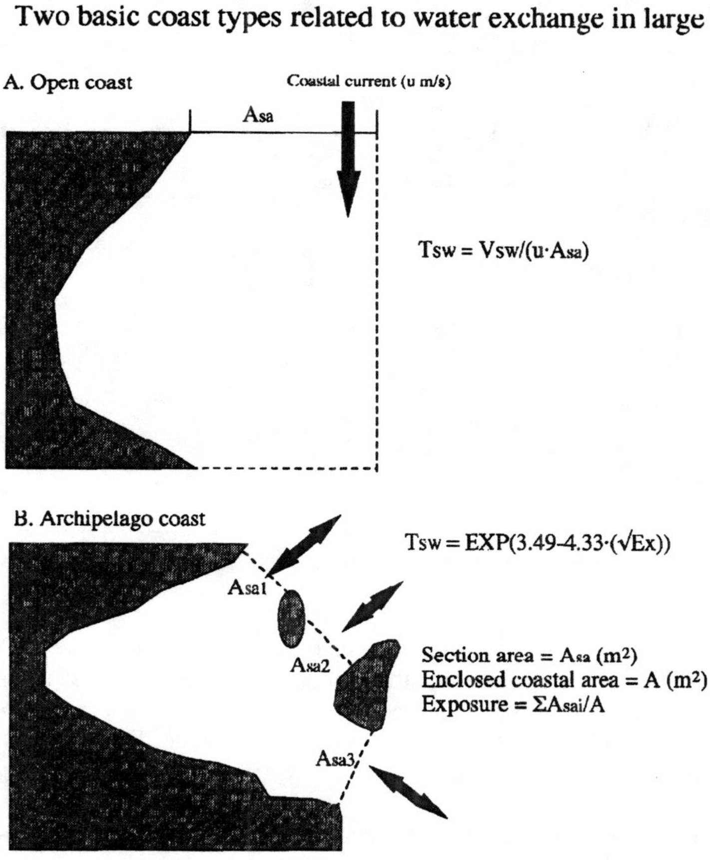

Fig. 2.6. Illustration of two main coast types (open coasts and closed archipelago coasts) in large lakes and how the theoretical surface water retention time (T_{SW}) may be estimated for each coast type.

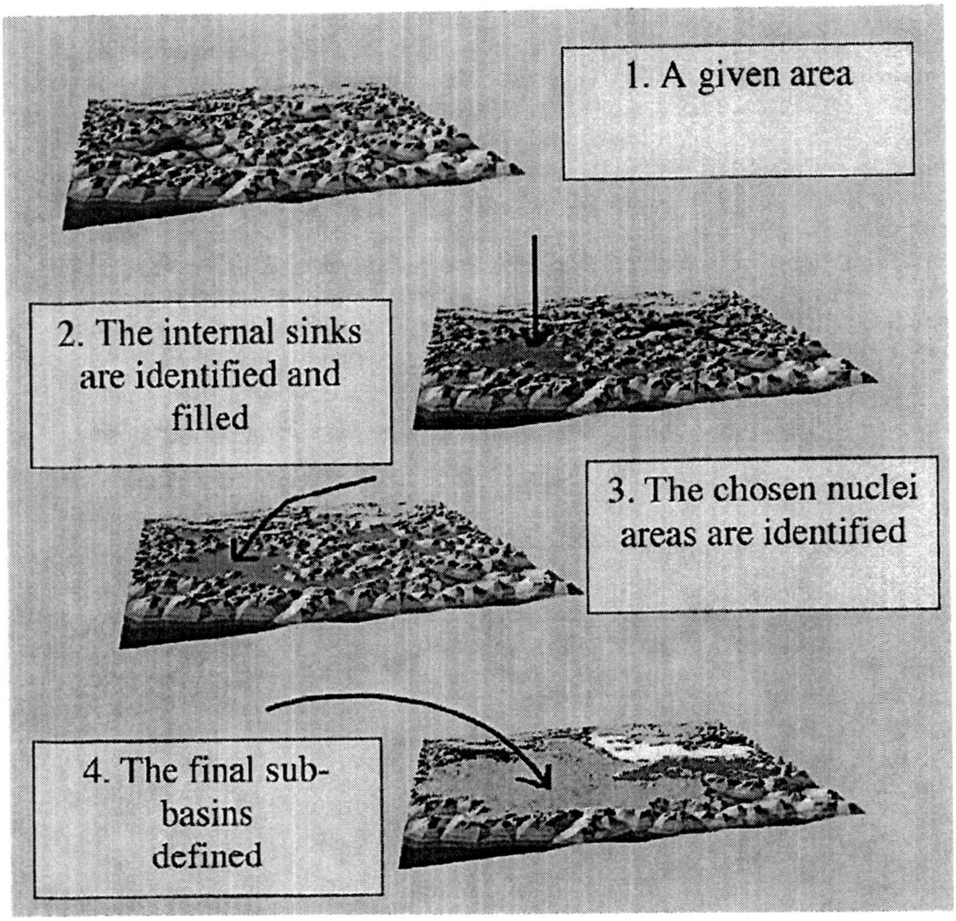

Fig. 2.7. Illustration of an automatic GIS method to define coastal boundary lines. From Lehtinen et al. (2001).

Fig. 2.7 illustrates a GIS approach (from Lehtinen et al., 2001; based on the ArcView program) to separate a given lake into topographically different basins. To make this separation may not be a major problem in small well-defined lakes, but it may be a major problem in large and topographically complex lakes; and there are many of those on Earth. This program will automatically search and identify sub-basins in the following steps:

1. First, all deep holes are identified. These may be key areas for a given sub-basin.
2. Then, the deep holes are filled in until they no longer constitute a hole.
3. The largest filled deep holes may then be selected as key areas.
4. Using a friction surface constructed from the basic topographical model, the key areas may grow until they meet.

The result will be a well-defined sub-basin; and for each sub-basin, one can easily calculate area, volume, mean depth, section area, etc., using the methods discussed in chapter 3. These morphometric data are, in turn, essential in mass-balance calculations and in effect-load-sensitivity analyses (see chapter 4).

3. Morphometric Parameters And Their Role In Lake Studies

Starting from the bathymetric map (see fig. 1.5), the aim of this chapter is (1) to define mathematically a set of lake morphometric parameters describing lake size, form and specific features and (2) to exemplify why and how these parameters are important in contexts related to the structure and function of lake ecosystems and in regulating the spread and effects of water pollutants. The latter aspects will be discussed more thoroughly in chapters 4 and 5.

Among the morphometric parameters, three main groups can be identified (see table 1.2 and fig. 3.1):

- Size parameters: (1) different parameters in length units, such as maximum length, water depth, shoreline length, maximum depth and maximum breadth, (2) parameters expressed in area units, such as 2D water surface area and 3D bottom area, and (3) parameters expressed in volume units, such as water volume and surface water volume (epilimnetic volume).
- Form parameters, based on size parameters, such as mean depth and shore development (= shoreline irregularity).
- Special parameters, like dynamic ratio, effective fetch and filter factor.

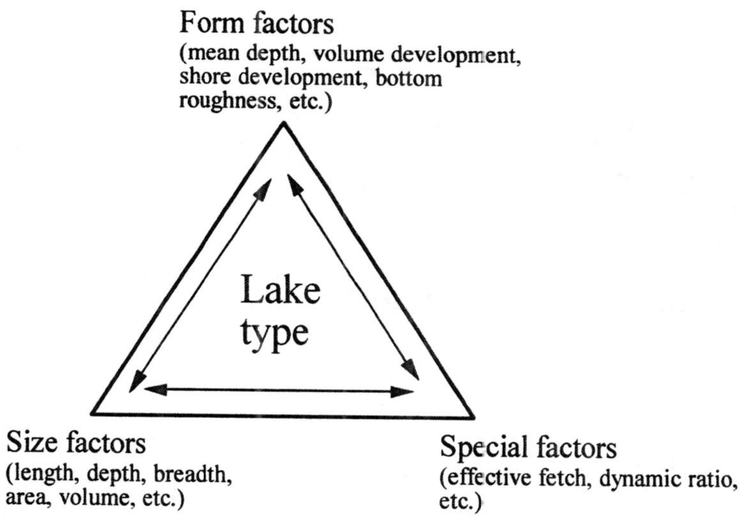

Fig. 3.1. The lake type may be expressed by morphometrical parameters describing different elements of size, form and special features.

This chapter will give an overview of how lake morphometrical parameters can be mathematically defined using both traditional methods (mostly from Håkanson, 1981a) and GIS approaches (mostly from Pilesjö et al., 1991).

3.1. Definitions of morphometrical parameters

3.1.1. Length measures

The maximum depth (D_{max} in m) is the greatest known depth in the lake. In GIS calculations, it is estimated by searching the raster data file. The minimum elevation value is equal to the maximum depth.

The maximum depth is very important in contexts of lake function, especially in defining the lake form factor (= volume development $V_d = 3 \cdot D_{mv}/D_{max}$; see later). Fig. 3.2A shows a frequency distribution for D_{max} based on data from 95 boreal glacial lakes (see appendix 8.3). One can note that the frequency distribution is positively skewed – the mean value (MV) is larger than the median (M_{50}). This skewing means that a logarithmic transformation (fig. 3.2B) will give a more normal frequency distribution (as given by the ratio between the mean value, MV, and the median value, M_{50}). This is information important in statistical analyses when one would like to test if variations in D_{max} among lakes influence variations in variables expressing functional aspects of lakes, for example, water clarity or primary production.

Fig. 3.3 illustrates that for these 95 lakes there is a strong, logical, positive correlation between maximum depth (D_{max}) and mean depth (D_{mv}) and that this correlation is stronger (the coefficient of determination, r^2, has increased from 0.80 to 0.88) if the regression is based on logarithmic data for both D_{max} and D_{mv}. The results mean that variations among these lakes in maximum depths can statistically explain 88% rather than 80% of the variations that exists among these lakes in mean depth values.

Chapter 4 will give many regressions like this between lake morphometric parameters and variables expressing important lake functional aspects, and appendix 8.1 gives a short compilation of basic statistical methods and concepts.

Traditionally, the mean depth (D_{mv} in m) is defined by the ratio between the lake volume (V in m^3) and the lake area (A in m^2), or $D_{mv} = V/A$. In GIS work, the mean depth may be estimated from raster data. The depths of all pixels with an elevation value equal to or below zero (water pixels) are summed, and the sum is divided by the number of pixels. To estimate the mean depth below a given depth, only water pixels with a depth equal to or deeper than the given depth are used in the calculations. Fig. 3.4 gives frequency distributions for mean depths from 259 lakes. Fig. 3.4A gives graphs for actual data and figures 3.4B and C show two transformations giving more normal frequency distributions, an exponential transformation, $D_{mv}^{0.1}$ and a logarithmic transformation, $\log(D_{mv})$.

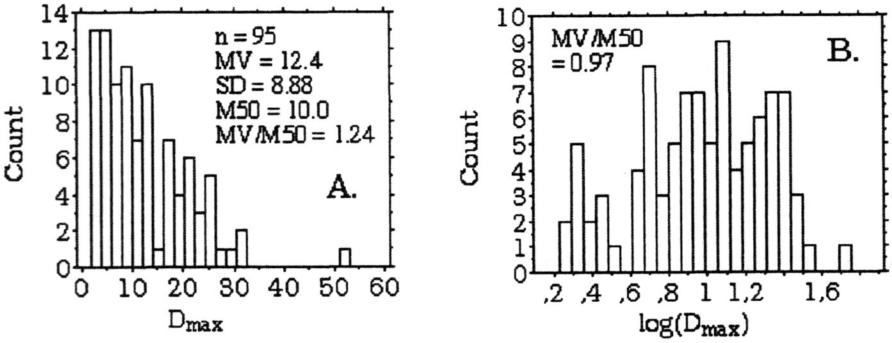

Fig. 3.2. Frequency distributions for the maximum depth (D_{max}) based on data from 95 boreal glacial lakes (data from Håkanson and Peters, 1995).

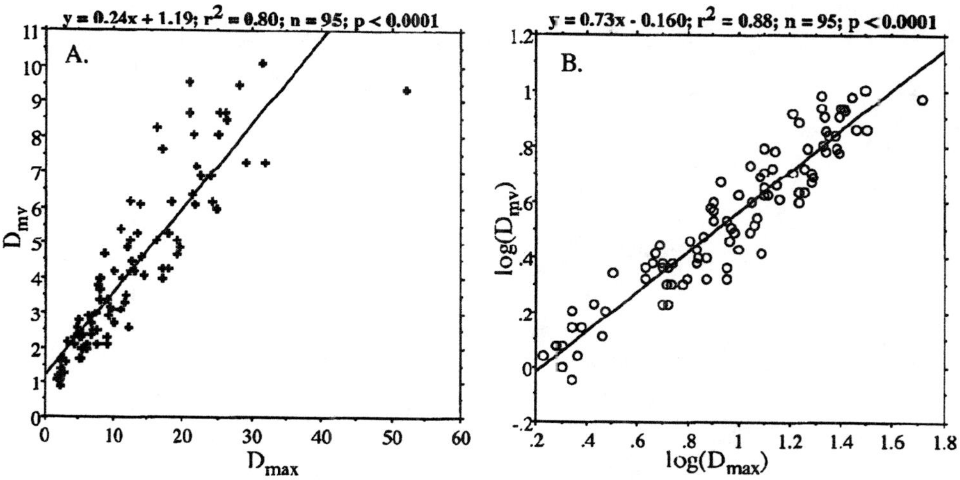

Fig. 3.3. Regression between maximum depth (D_{max}) and mean depth (D_{mv}) using actual data (A) and logarithmic data (B) for 95 boreal glacial lakes. The figure gives the regression line, the coefficient of determination (r^2), the number of data (n) and the statistical certainty (p). Data from Håkanson and Peters (1995).

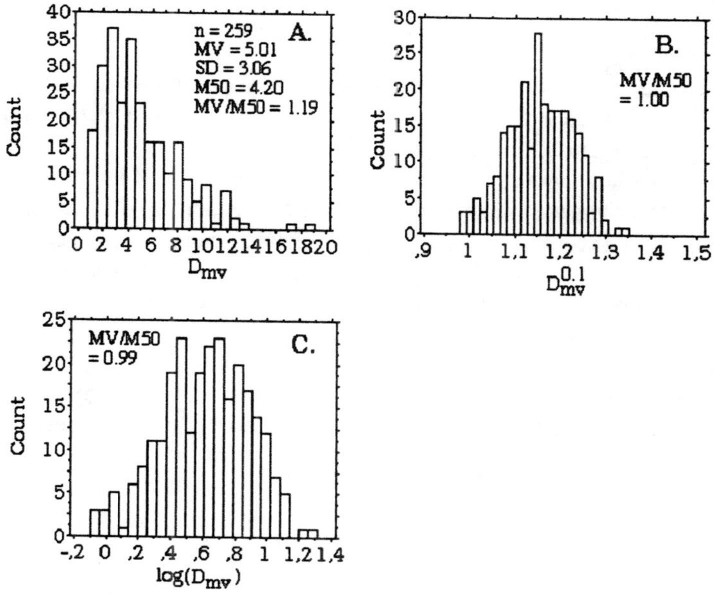

Fig. 3.4. Frequency distributions for the mean depth (D_{mv}) based on data from 259 lakes (data from Håkanson and Peters, 1995).

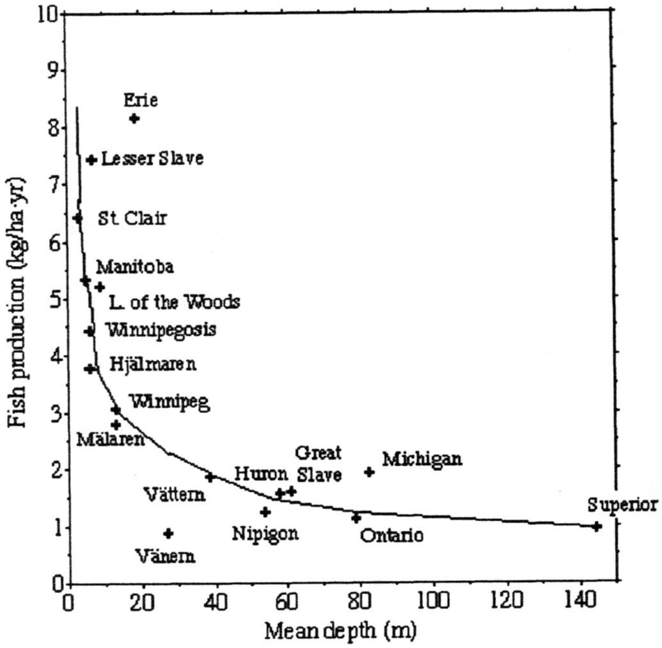

Fig. 3.5. The relationship between commercial fish production and lake mean depth (data from Rawson, 1955 and Ahl, 1975).

The mean depth is a most informative and useful parameter in many contexts in limnology. This is exemplified here from the relationship between mean depth and (1) commercial fish production (fig. 3.5) and (2) Secchi depth (fig. 3.6).

The latter is a key variable of water clarity and the depth of the photic zone, and hence also a key regulator of primary and secondary production. From fig. 3.6, one can note that 46% of the variations in mean Secchi depths among these 88 lakes can be explained statistically by variations in mean depth. Also note that this regression uses the long-term (three year mean value of Secchi depth). Short-term variations in Secchi depth are generally related to variations in water temperature and light conditions.

The median depth (D_{50} in m); 50% of the lake area lies below the D_{50} value and 50% above. The median depth is traditionally determined from the percentage hypsographic curve (fig. 3.7). D_{50} is 22 m in the example in fig. 3.7 using data for Lake Vänern, Sweden.

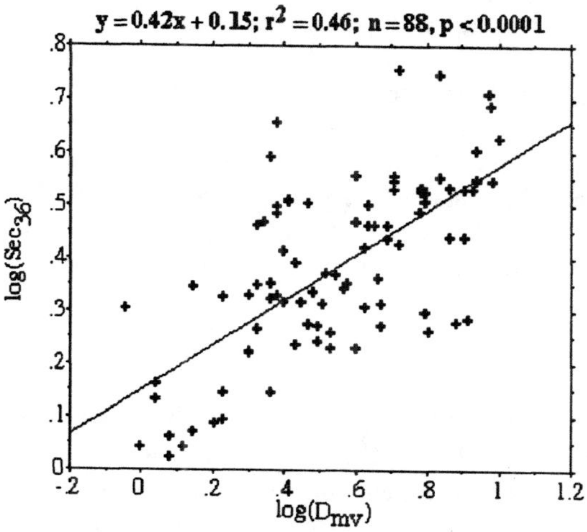

Fig. 3.6. The relationship between the mean depth [log(D_{mv})] and Secchi depth (logarithmic transformation of the mean value based on data from 36 months for 88 lakes; data from Håkanson and Peters, 1995).

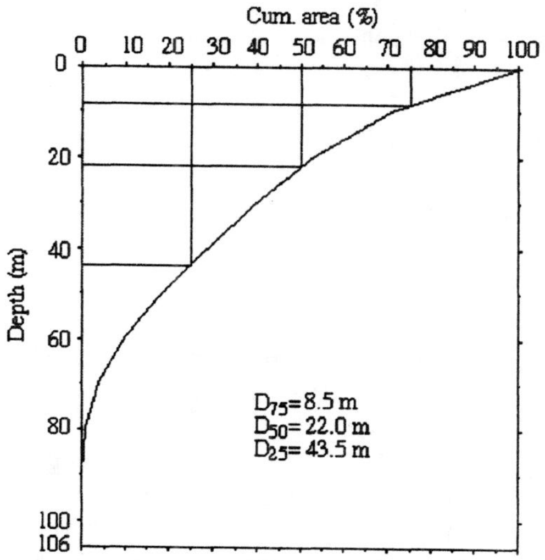

Fig. 3.7. Determination of the median depth (D_{50}) and the quartile depths (D_{25} and D_{75}) from the percentage hypsographic curve for Lake Vänern.

In lakes with convex relative hypsographic curves (see fig. 3.48), like Lake Vänern, the mean depth, D_{mv}, is larger than the median depth, D_{50}. In lakes with concave relative hypsograhic curves, the opposite is true.

In GIS approaches, D_{50} may be estimated from the raster data. First, the number of pixels with depths between zero and one meter, one and two meters and so on (up to the class: integer of the maximum depth minus integer of the maximum depth minus 1) are calculated. The results from this calculation may be compared to data in a frequency table, where the frequencies correspond to the number of pixels and the class boundaries are the upper and lower depths for the intervals. Then, the median depth may be calculated from:

$$D_{50} = u+(n/2-c)/f \qquad (3.1)$$

Where n is the number of water pixels in the area, u is the lower class boundary in the class containing the $n/2^{th}$ pixel (ranked by depth), f is the frequency in that class and c is the number of pixels below that class. To estimate the median depth below a given depth, only water pixels with a depth equal to or deeper than this depth are used in the calculations.

The median depth may be used, for example, to determine the lake bottom roughness (see eq. 3.23), which is a useful parameter in sedimentological contexts and in the optimization model for lake hydrographic surveys (see chapter 2).

The quartile depths (D_{25} and D_{75} in m) or the depth related to any given percentile (such as D_{10} or D_{90}) are traditionally also determined from the percentage hypsographic curve, as shown in fig. 3.7. Twenty-five percent of the lake area should lie below and 75% above the depth given by the first quartile, D_{25}. The median depth may also be called the second quartile depth. The values for Lake Vänern are D_{25} = 43.5 m and D_{75} = 8.5 m.

Using GIS, these depths are estimated in a similar manner to that used for the median depth. Eq. 3.1 is changed to:

$$D_{25} = u+(n/4-c)/f \qquad (3.2)$$

$$D_{75} = n+(3\cdot n/4-c)/f \qquad (3.3)$$

Where u is the lower class boundary in the class that contains the $n/4^{th}$ and the $3\cdot n/4^{th}$ pixel (ranked by depth) respectively.

To estimate the quartile depth below a given depth, only water pixels with a depth equal to or deeper than this depth are used in the calculations. The quartile depths are primarily to be considered as descriptive morphometrical parameters.

The maximum length (L_{max} in m; see fig. 3.8 for illustration using a map of Lake Vänern) is defined by the line connecting the two most remote points on the shoreline. In regular basins, L_{max} is generally straight and concurs with the maximum effective length (L_{xel}). In irregular lakes, such as oxbow lakes, L_{max} is a curved line. It may not cross land, but it may cross islands. This rule means that L_{max} cannot always be given a definite value. It is primarily to be considered as a descriptive measure and has limited use in contexts on lake function. The L_{max} value is 140 km in Lake Vänern, Sweden.

The direction of major axis is defined by the general compass direction of the maximum length. It is SW-NE in Lake Vänern (see fig. 3.8)

Maximum effective length (L_{xel} in m; see fig. 3.8) is defined by the straight line connecting the two most distant points on the shoreline over which winds and waves may act without interruptions from islands and land. This parameter is an important one in many lake studies, such as those concerning internal seiches (see fig. 3.9).

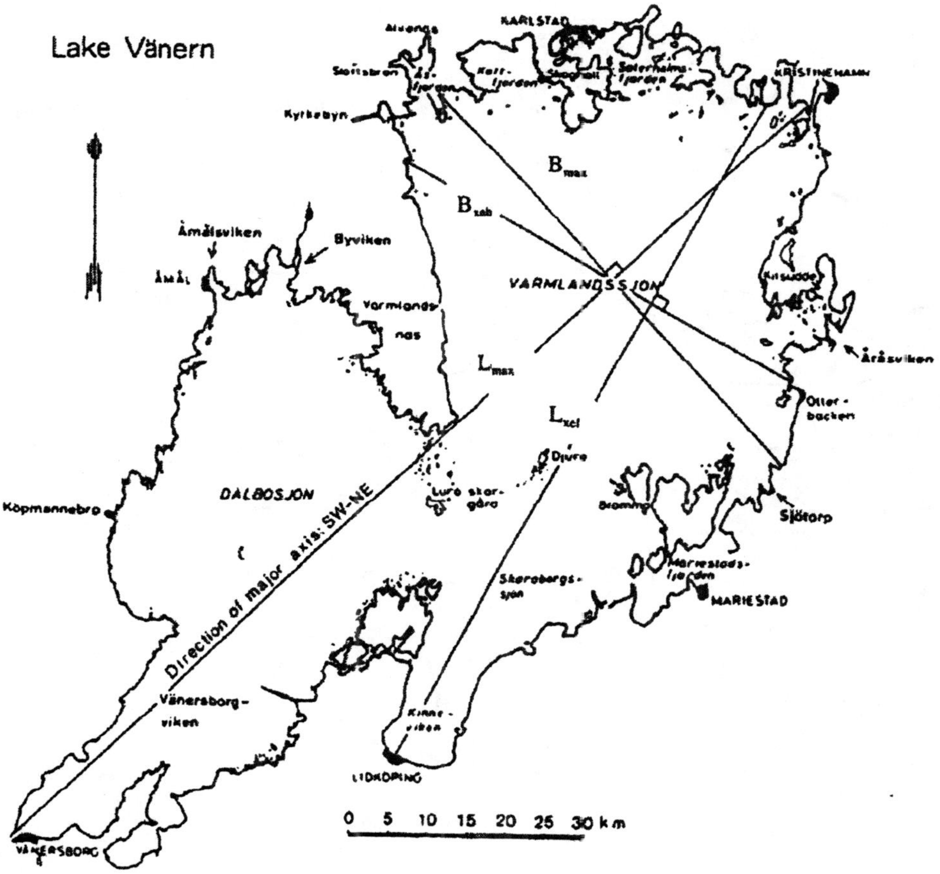

Fig. 3.8. Orientation map for Lake Vänern and illustration of maximum length (L_{max}), maximum effective length (L_{xel}), maximum breadth (B_{max}), maximum effective breadth (B_{xeb}) and direction of major axis.

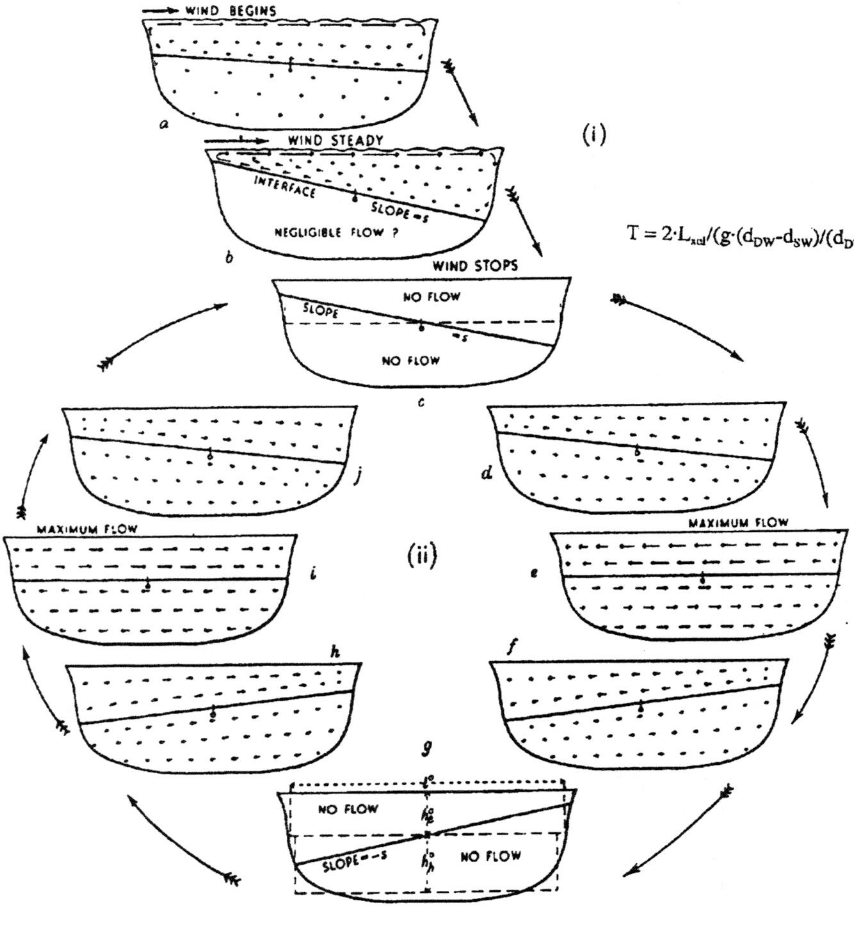

$$T = 2 \cdot L_{xel}/(g \cdot (d_{DW} - d_{SW})/(d_{DW}/D_{DW} + d_{SW}/D_{SW}))^{0.5}$$

Fig. 3.9. Water movements caused by wind stress and subsequent internal seiche activity in a stratified hypothetical lake without friction. Direction and velocity of the flow are indicated by arrows. t = period of internal seiche, L_{xel} = maximum effective length, g = acceleration due to gravity (980.6 cm/s), d_{SW} = density of surface water, d_{DW} = density of deep water, D_{SW} = depth of surface water and D_{DW} = depth of deep water layer (from Mortimer, 1952).

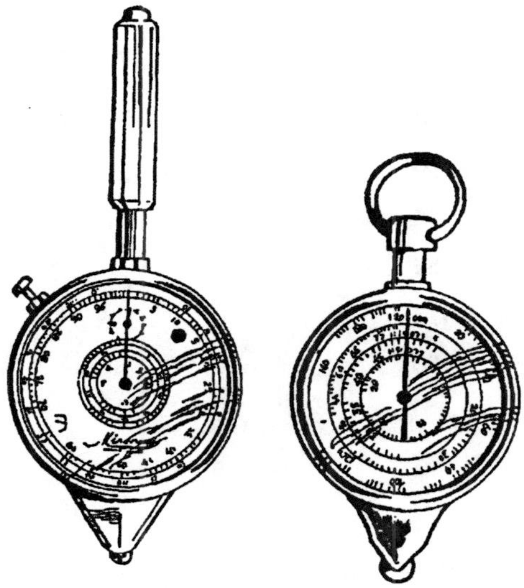

Fig. 3.10. Two types of map measurers.

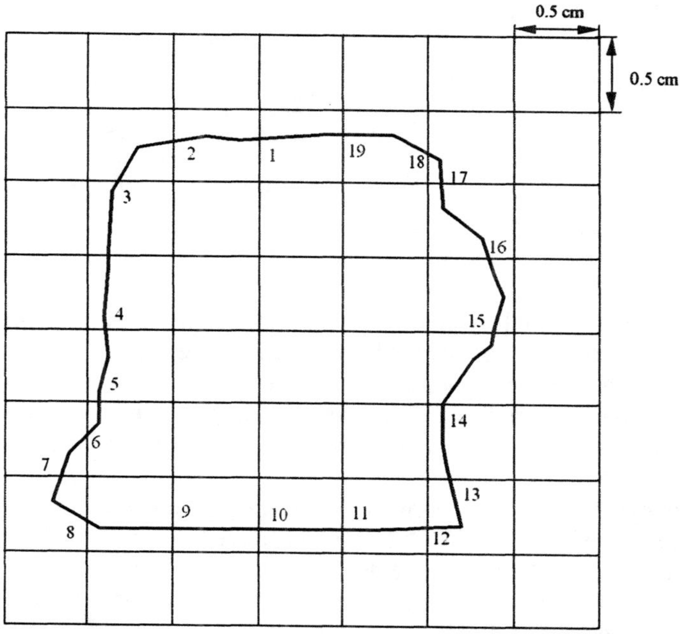

Fig. 3.11. Illustration of the CTP technique for determination of the total shoreline length (L_{tot}). The number of intersectional points in this example is x = 19.

Table 3.1. The relationship between the number of necessary turns (n') to obtain the shoreline length with a given statistical certainty and the number of intersectional points (x) obtained with the CTP technique.

X	n'
>475	1
474-350	2
349-260	3
259-205	4
204-160	5
159-125	6
124-100	7
99-82	8
81-70	9
69-59	10
58-51	11
50-46	12
45-42	13
41-39	14
38-36	15
35-34	16
33	17
32-31	18
30	19
29	20

Traditionally, the shoreline length (L_0, in m) may be estimated in different ways, for example, (1) by map measurer (see fig. 3.10) and (2) by dot counting (see fig. 3.11 and table 3.1).

The latter approach is explained in more detail below:

1. The total lake area (A_{tot}), i.e., the water surface area (A) plus the total areas of islands (A_{itot}) is assumed to be known. $A_{tot} = 5893$ km^2 for Lake Vänern.

2. The optimal map scale factor (SF) for the determination of the shoreline length is given by: $SF \approx 5500 \cdot \sqrt{A_{tot}}$. That is, for Lake Vänern, $SF = 5500 \cdot \sqrt{5893} \approx 422{,}000$. The best (the largest available) map scale that corresponds to this scale factor should be used. For Lake Vänern, there exist possible maps in scales 1:250,000 (best) and 1:500,000.

3. A checkered transparent paper (CTP), having a square size of 0.5 cm, is then put over the map and the number of intersectional points (x) between the shoreline and

the squares counted (see fig. 3.11). Using the map in scale 1:250,000, a total of 1286 intersectional points were obtained. No further determinations are then required to establish the shoreline length with a given statistical confidence (the error in the length determination should, with a 95% certainty, be less than 5%). If one had used another map, for example, in scale 1:1000,000, one would have obtained a different x-value, say x = 200. To obtain the shoreline length with the same statistical certainty, one must then make more determinations. The relationship between the number of necessary turns or determinations (n' = the number of times the CTP paper is put on the map and the x-value determined) and the number of intersectional points (x) is given in table 3.1. From this table, one can see that for n = 200, n' is 5, and four more determinations should be done. The mean value (x_{mv}) of those five measurements is 202.

4. The <u>scale-dependent shoreline length</u> (L_0' in km) can be calculated from:

$$L_0' = 0.5 \cdot x_{mv} \cdot SF \cdot 10^{-5} \qquad (3.4)$$

That is, for Lake Vänern using the map scale 1:250,000, one gets:

$L_0' = 0.5 \cdot 1286 \cdot 250000 \cdot 10\text{-}5 = 1608$ km

and from the map in scale 1:1,000,000, one has:

$0.5 \cdot 202 \cdot 1000000 \cdot 10\text{-}5 = 1012$ km.

Evidently, the result depends on the map scale: the better the map (i.e., the smaller the scale factor), the better the value of the shoreline length.

The requested <u>normalized scale-independent value of the shoreline length</u> (L_0) may then be calculated from the normalization formula (Håkanson, 1978b):

$$L_0 = L_0' \cdot (K_2 - K_1)/(K_2 - \log(SF + a')) \qquad (3.5)$$

Where

a'	$= 10^5 \cdot \log(A_{tot})$, 10^5 = the area constant,
A	= the total lake area in km^2 (not m^2);
a'	$= 10^5 \cdot \log(5893) = 377034$ for Lake Vänern;
SF	= the scale factor (= 1/MS; MS = the map scale);
K_1	= log(SF + a') for SF = 1, i.e., the reference scale (1:1);
K_1	= 5.576 for Lake Vänern;
K_2	= log(SF + a') for SF = 6,000,000, where 6,000,000 is the scale constant (an operational maximum scale factor);
K_2	= 6.805 for Lake Vänern.

Then the data for Lake Vänern (scale 1:250,000) gives:

$L_0 = 1608 \cdot (6.805 - 5.576)/(6.805 - \log(250000+377034)) = 1608 \cdot 1.22 = 1960$ km.

The data from the map in scale 1,000,000 gives in the same way L_0 = 1870 km. The difference between those two values is 90 km or 4.6%, which is within the acceptable statistical limits. Since the value 1960 km is obtained from a better map, this value is preferable. The mean value from many calculations like these for Lake Vänern is 1943 km.

Using GIS approaches, the shoreline length (L_0 in km) may be estimated from the vector data, where all coordinates for the coastline are stored. The formula

$$L_0 = \sum\sqrt{((x_i - x_i + 1)^2 + (y_i - y_i + 1)^2)} \qquad (3.6)$$

gives the shoreline length when i ranges between 1 and (n-1). In order to make L_0 scale-independent, a limit for the distance between two consecutive coordinates has to be set. The limit may be set to 50 meters, which implies that coordinates digitized with a distance less than 50 meters are eliminated from the calculation. This elimination makes it possible to compare length measurements obtained from charts with different map scales.

The <u>shoreline length of an island</u> (L_{is} in km) is also estimated from eq. 3.6 using GIS and by map measurer or CTP by traditional methods and the <u>total shoreline length</u> (L_{ist} in km) is estimated by summation of these islands.

To know the shoreline length is important to determine the shore development (L_d) and to calculate production and/or biomass of benthic algae and macrophytes (see Kalff, 2002).

Normalized scale-independent values of the <u>contour line length</u> (L_{ci} in km) at a defined water depth (i) may also be estimated by map measurer or CTP using traditional methods or by eq. 3.6 using GIS and the total length of all contour lines (L_{cit}) as the sum of these individual contour lines.

Values of contour line length are used to calculate slope and bottom roughness, among other things.

<u>Maximum breadth or width</u> (B_{max} in km) is defined by the straight line at the right angle to the maximum length (L_{max}) that connects the two most remote extremities on the shoreline without crossing land. Islands may be crossed. B_{max} has, like L_{max}, primarily a descriptive value in limnology. B_{max} is 72 km in Lake Vänern (see fig. 3.8).

The <u>maximum effective breadth</u> (B_{xeb} in km) is defined by the straight line on the lake surface perpendicular to the maximum effective length, L_{xel}, which connects the two most distant sites on the shoreline. Thus, B_{xeb} may not cross land or islands. B_{xeb} is 57 km in Lake Vänern (see fig. 3.8).

The <u>mean breadth</u> (B_{mv} in km) is defined by the ratio between the lake area (A in km^2) and the maximum length (L_{max} in km), i.e., $B_{mv} = A/L_{max}$.

The mean breadth is 40 km in Lake Vänern. The maximum effective breadth and the mean breadth are, like the maximum effective length, useful parameters in lake hydrodynamics.

3.1.2. Measures of area and volume

Traditionally, <u>total lake area</u> (A_{tot} in m^2) is calculated by planimeter (see fig. 3.12) using a map in a given scale. If the scale factor (SF) is increased, small caps and bays will disappear on the map. This change will primarily affect determinations of shoreline length and not the area determination, since planimetration implies that negative errors (bays) are leveled out by positive errors (caps), provided that the planimetration is carried out on a map of reasonable scale, which is given by $SF \leq 5500 \cdot \sqrt{A_{tot}}$. So, one can regard the area determination as rather independent of the map scale.

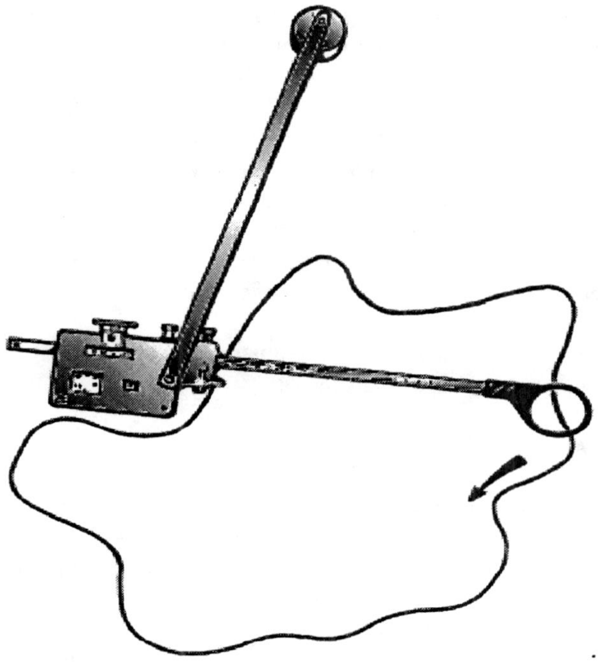

Fig. 3.12. The ARISTO planimeter.

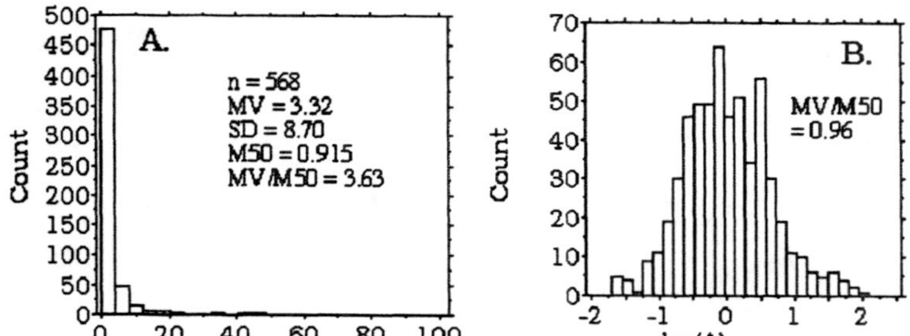

Fig. 3.13. Frequency distributions for lake area and for log(A) based on data from 568 lakes (data from Håkanson and Peters, 1995).

Figure 3.13 gives frequency distributions for lake area for 568 lakes; and one can note that there are in this sample, and on this planet, many more small lakes than large lakes. A logarithmic transformation gives a frequency distribution that is close to normal (fig. 3.13B). The total area of our reference lake, Lake Vänern, is as already stated 5893 km^2.

When using GIS to determine lake area, the coordinates describing the shoreline are stored in a vector file. The total area is estimated by:

$$A_{tot} = ABS(\sum(x_i \cdot y_i + 1) - \sum(y_i \cdot x_i + 1))/2.0 \qquad (3.7)$$

Where x_i and x_{i+1} are the x-coordinates for two consecutive points, y_i and y_{i+1} are the corresponding y-coordinates, and i ranges between 1 and n. Point (n+l) is identical to point 1 (the polygon is closed). ABS is the symbol for absolute value.

Estimations of areas according to eq. 3.7 are also scale-independent since "positive errors" (bays) are compensated for by "negative errors" (points of land).

The total area (A_{tot}) is an important parameter in many limnological contexts, but the water surface area (A in m^2) may be even more important. The water surface (A) is given by the difference in total area (A_{tot}) and the total area of islands, islets and rocks (A_{ist}). Using GIS, A is calculated according to eq. 3.7. To estimate the water area at, for example, the water depth of the wave base (D_{wb}), one can use raster data. The areas of each pixel with a depth equal to or less than D_{wb} in the lake are then summed.

The 3D bottom area (A_{3D}, m^2) may be estimated using GIS by dividing the water surface area (A) by the cosine of the mean slope in radians (S_{mv}):

$$A_{3D} = A/\cos(\mathrm{atan}(S_{mv}/100)) \qquad (3.8)$$

If one would like to estimate the bottom area below a given depth (such as D_{wb}), the water area is calculated from eq. 3.7; and pixels with depths less than D_{wb} are excluded from the mean slope estimation. The bottom area above D_{wb} is estimated by subtracting the bottom area below D_{wb} from the total bottom area.

Using GIS, the areas of single islands (A_{is} in m^2) are estimated according to eq. 3.7. The areas of single islands at given depths are estimated by summation of the pixel areas of an island at that depth in the lake.

The total area of all islands (A_{ist} in m^2) is estimated by summation of all the areas of the islands.

Using traditional methods, the lake water volume (V in m^3) may be determined from the following two formulae:

$$V_l = \sum_{i=o}^{n} (L_c/2)\cdot(A_i + A_i+1) \qquad (3.9)$$

$$V_p = \sum_{i=o}^{n} (L_c/3)\cdot(A_i + A_i+1 + \sqrt{(A_i + A_i+1)}) \qquad (3.10)$$

Where

L_c = The contour line interval (m).
A_i = The total area within the limits of the contour line L_i in km^2.

The V_l formula (eq. 3.9) is called the linear approximation of the volume, the V_p formula (eq. 3.10), the parabolic approximation. Which of these two formulae should be used depends on the lake form, as given by the relative hypsographic curve (see fig. 3.14 for Lake Vänern). The V_l formula provides best values for lakes with concave relative hypsographic curves, the V_p expression is preferable for lakes with convex relative hypsographic curves. The calculation routine for lake volume will be exemplified here for Lake Vänern. Basic data on depth and area are given in table 3.2. Lake Vänern has a form that is classified as convex (= Cx, see fig. 3.14), which means that the relative hypsographic curve lies between the limitation lines f(-1.5) and f(-0.5). These limitation lines f(-1.5) and f(-0.5) represent 1.5 and -0.5 standard deviations (SD) from the mean lake form, f(MV). From basic statistics, one can note that there is a 70% chance that data in a normal frequency distribution will appear within ± one standard deviation from the mean and a 95% chance that data will appear within ± two standard deviations from the mean.

Table 3.2. Lake Vänern, data on depth, area and volume.

Depth (m)	Cumulative depth (%)	Area (km^2)	Cumulative area (km^2)	Cumulative area (%)	Volume (km^3)	Cumulative volume (km^3)
0-10	0	1653	5648.4	100	47.981	153.45
10-20	9.43	1038	3995.4	70.7	34.634	105.47
20-30	18.9	723.7	2957.4	52.4	25.871	70.836
30-40	28.3	629.8	2233.7	39.5	19.101	44.965
40-50	37.7	578.0	1603.9	28.4	13.042	25.864
50-60	47.2	481.0	1025.9	18.2	7.728	12.823
60-70	56.6	335.9	544.87	9.6	3.638	5.095
70-80	66.0	160.0	208.97	3.70	1.197	1.457
80-90	75.5	43.6	48.97	0.87	0.235	0.260
90-100	84.9	5.00	5.37	0.10	0.024	0.025
100-106	94.3	0.37	0.37	0.01	0.001	0.001
106	100	0	0	0	0	0

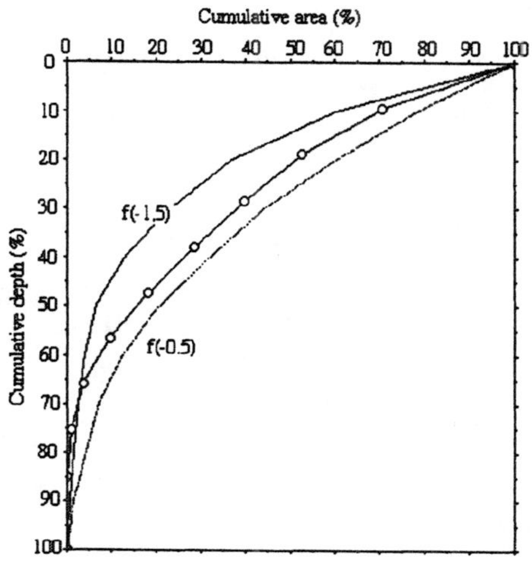

Fig. 3.14. The relative hypsographic curve for Lake Vänern between the limitation lines f(-1.5) and f(-0.5).

The uncorrected volume according to the V_p formula is 153.45 km^3 for Lake Vänern. This value can be improved so that the uncertainty in the volume is minimized. The correction factor is obtained from table 3.3. Using a bathymetric map with 10 contour lines (n = 10) for a lake with a convex (Cx) lake form, one gets a correction factor of 0.996. That is, the corrected volume for Lake Vänern is 0.996·153.45 = 152.84 km^3. The true volume should lie between the limits obtained when the correction factor for the limitation lines f(-1.5) and f(-0.5) are multiplied with the uncorrected V_p value of 153.45 km^3. The correction factor for the f(-1.5) line for n = 10 is 0.990 (see table 3.4); and for the f(-0.5) line for n = 10, it is 0.999. That is, the true volume should lie between 153.30 km^3 (0.999·153.45) and 151.92 km^3 (0.990·153.45). The maximum percentage error in the corrected volume 152.84 is between +0.3% and –0.6%.

Table 3.3. Correction factor table for various lake forms and numbers of contour lines in the bathymetric map (n).

		n=2	n=3	n=4	n=5	n=6	n=8	n=10
V_p	VCx	0.627	0.799	0.878	0.916	0.945	0.966	0.977
	Cx	0.930	0.969	0.981	0.988	0.990	0.993	0.996
	SCx	1.033	1.017	1.010	1.005	1.004	1.003	1.002
V_l	L	0.984	0.992	0.997	0.999	0.999	1.000	1.000
	C	1.046	1.023	1.012	1.007	1.006	1.003	1.002

Table 3.4. Correction factors for various relative hypsographic curves in the classification system for lake forms.

n	f(-3)	f(-2)	f(-1.5)	f(-1)	f(-0.5)	f(MV)
1	0	0	0.270	0.661	0.915	1.107
2	0	0.539	0.788	0.912	0.989	1.033
5	0.358	0.893	0.960	0.985	0.997	1.005
10	0.747	0.971	0.990	0.994	0.999	1.002
25	0.944	0.994	0.999	1	1	1
50	0.995	0.999	1	1	1	1
n	**f(MV)**	**f(0.5)**	**f(1)**	**f(1.5)**	**f(2)**	**f(3)**
1	0.660	0.865	1.021	1.156	1.261	1.404
2	0.920	0.959	0.991	1.021	1.059	1.135
5	0.987	0.997	0.997	1.002	1.010	1.038
10	0.997	0.999	1	1	1.002	1.011
25	0.999	1	1	1	1	1.002
50	1	1	1	1	1	1

This method to determine the lake volume is, in principle, valid for all lake types. For lakes with rather more complex relative hypsographic curves, however, the correction factor may be established in a more detailed way. This procedure will be illustrated using data from Lake Great-Hjälmaren, Sweden. Data on depth and area for this lake are given in table 3.5. There are seven contour lines (n = 7) in this case. Fig. 3.15 shows that 28% of the area falls within the limits for convex lake form (Cx), that is, between the lines f(-1.5) and f(-0.5); 72% of the area is within the SCx class (slightly convex) given by the lines f(-0.5) and f(0.5). Thus, the SCx class is the main class, implying that the V_p formula should be used in the volume determination. The uncorrected volume is 2.001 km^3. A correction factor may be established using the data in table 3.3. For the Cx class for n = 7, one has the correction factor (0.990+0.993)/2 = 0.9915; for the SCx class, it is (1.003+1.004)/2 = 1.0035. A correction factor may then be calculated from: 0.28·0.9915+0.72·1.0035 = 1.00014. The corrected volume of Lake Great-Hjälmaren is 2.001·1.00014 = 2.0013 km^3.

Table 3.5. Lake Great-Hjälmaren data (from Håkanson, 1981a) on depth, area and volume; n = contour line number

Depth (m)	Cumulative depth (%)	Area (km^2)	Cumulative area (km^2)	Cumulative area (%)	Volume (km^3)	Cumulative volume (km^3)	n
0-3	0	56.15	277.5	100	0.747	2.001	
3-6	13.6	44.36	221.3	79.8	0.596	1.254	1
6-9	27.3	88.55	177.0	63.8	0.390	0.658	2
9-12	40.9	54.88	88.42	31.9	0.176	0.268	3
12-15	54.5	9.7	33.54	12.1	0.069	0.091	4
15-18	68.2	12.2	13.87	5.00	0.020	0.022	5
18-21	81.8	1.69	1.71	0.62	0.002	0.002	6
21-22	95.5	0.02	0.02	0.01	0.00002	0.00002	7
22	100	0	0	0	0	0	

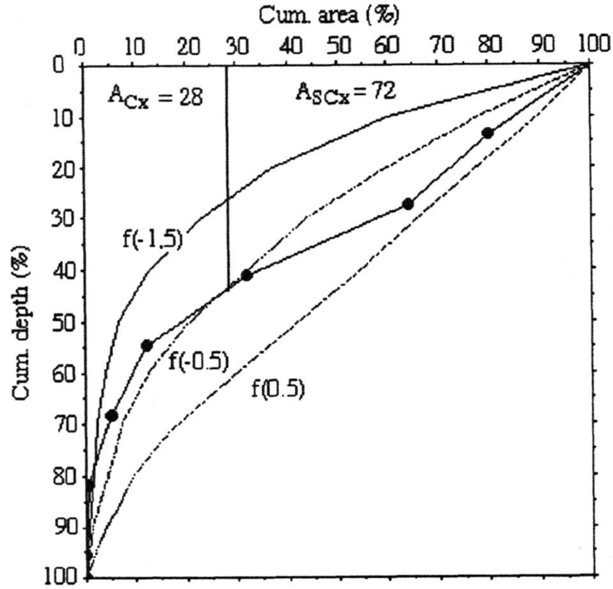

Fig. 3.15. The relative hypsographic curve for Lake Great-Hjälmaren (data from Håkanson, 1981a).

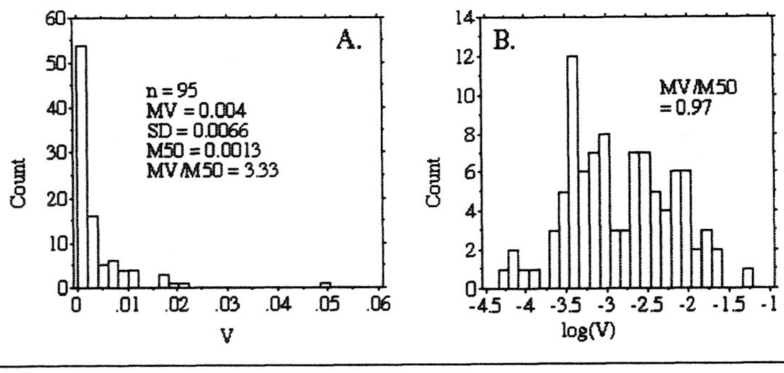

Fig. 3.16. Frequency distributions for lake volume (V) and log(V) based on data from 95 boreal glacial lakes (data from Håkanson and Peters, 1995).

Figure 3.16 gives a frequency distribution for lake volume for 95 glacial boreal lakes. One can note that there are many more small lakes than large lakes, so the frequency distribution is positively skewed and a logarithmic transformation (fig. 3.16B) will produce a more normal frequency distribution.

Using GIS, the water volume is often estimated by multiplication of the water surface area (A) by the mean depth (D_{mv}). To estimate the water volume above or below a given depth, the raster data file is often used. The area of each pixel is multiplied by the distance to the given depth and summed.

The lake volume is a fundamental lake variable needed to calculate, for example, lake concentrations of all types of substances (C = M/V) and the theoretical water retention time ($T = V/Q_{mv}$; Q_{mv} is the mean tributary water discharge) regulating the retention time of all substances in lakes.

3.1.3. Special measures and measures of form

3.1.3.1. Different parameters related to measures of length

The relative depth (D_{rel}) is defined by the ratio between the maximum depth (D_{max}) and the mean diameter of the lake (using the lake water area, A):

$$D_{rel} = (D_{max} \cdot \sqrt{\pi})/(20.0 \cdot \sqrt{A}) \qquad (3.11)$$

The relative depth is 0.12% in Lake Vänern, which is a normal figure for large basins. Small and deep lakes have high D_{rel} values (see fig. 3.17). The relative depth is often used as a measurement of the stability and stratification of water masses and to predict oxygen conditions in lakes (Eberly, 1964).

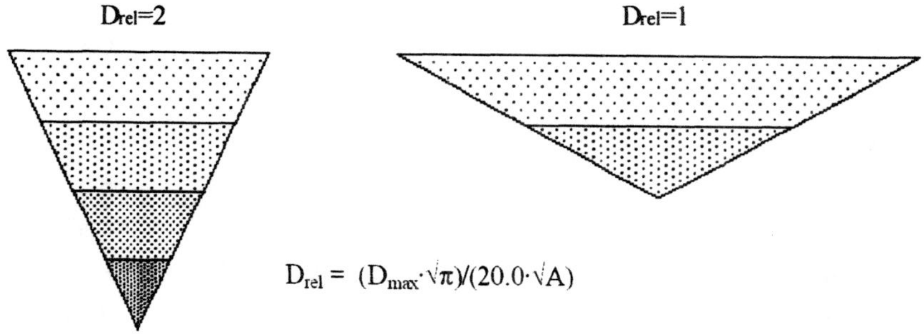

Fig. 3.17. Illustration of the relative depth (D_{rel}) for two different lakes

The mean breadth of coastal areas (in large lakes; B_{cmv}, in m or km) is defined by the ratio between the total coastal area (A_c) and the shoreline length of the given coastal area (L_{co}):

$$B_{cmv} = A_c/L_{co} \qquad (3.12)$$

Large coastal areas with high shore irregularity (see below) have high B_{cmv} values while small areas with low shore irregularity have low B_{cmv} values.

Traditionally, the shore development (shoreline irregularity, L_d) is a dimensionless measure of the degree of shoreline irregularity and defined by:

$$L_d = L_0/2 \cdot \sqrt{(\pi \cdot A_{tot})} \qquad (3.13)$$

Where L_0 is the normalized shoreline length (in km) and A_{tot} is the total lake area (in km^2). The L_d value illustrates the relationship between the actual length of the shoreline and the length of the circumference of a circle with an area equal to the total lake area. A perfectly circular basin has a L_d value of 1. Irregular basins have higher L_d values. L_d values larger than 10 are rare, and one of the probably most irregular basins in the world, Lake Mälaren, Sweden, has an L_d value of 9.95. This lake is basically a big archipelago with many communication basins and bays.

It should be stressed again that all scale-dependent parameters, like the shoreline length, the shore development and the contour lines length, have limited quantitative meaning if they are not normalized. The relationship between map scale, shoreline length (L_0) and shore development

(L_d) is illustrated in table 3.6 using data for Lake Vänern. The shoreline length and the shore development differ by about 100% if determined on maps with large scale factors compared to the normalized value of 1943 km for L_0 and 7.14 for L_d.

Using GIS, the shore development may be determined by the ratio L_0/L_{min}, where L_0 is the shoreline length and L_{min} a measure of the shortest distance, with the same integral (Int) as the integral below the shoreline, between the two end points, a and b (see fig. 3.18).

Table 3.6. The relationship between map scale, shoreline length and shore development for Lake Vänern (data from Håkanson, 1981a).

Scale factor (SF)	**Shoreline length** (L_0, km)	**Shore development** (L_d)
10,000	2007	7.38
50,000	1875	6.89
60,000	1893	6.96
180,000	1642	6.03
250,000	1608	6.03
500,000	1325	4.87
1,000,000	1012	3.72
Normalized	1943	7.14

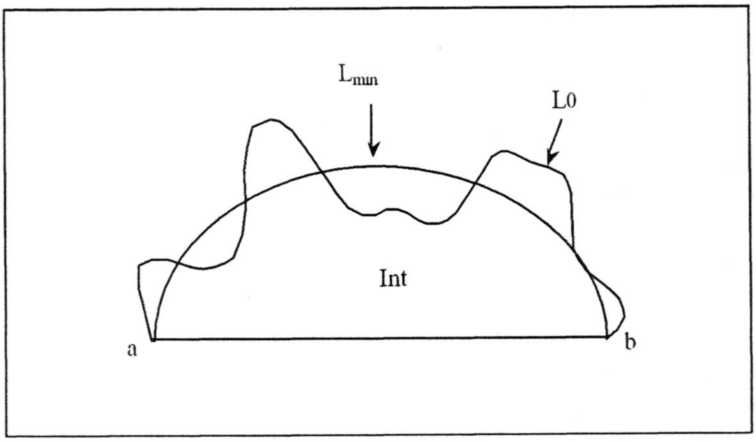

Fig. 3.18. Illustration of a GIS expression for the shore development ($L_d = L_0/L_{min}$).

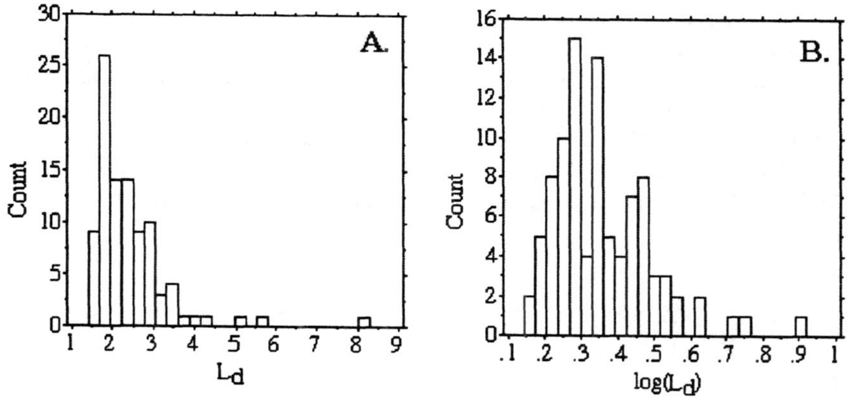

Fig. 3.19. Frequency distributions for the shore development (L_d) and log(L_d) based on data from 95 boreal glacial lakes (data from Håkanson and Peters, 1995).

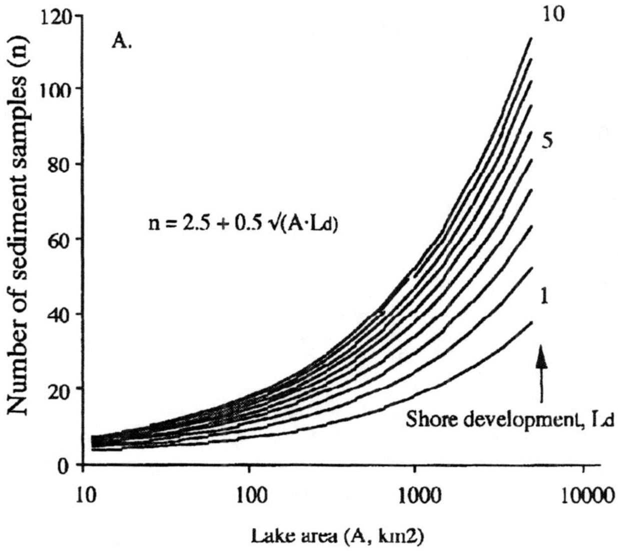

Fig. 3.20. Nomogram illustrating the formula for sediment sampling (for more information, see Håkanson and Jansson, 1983).

Fig. 3.19 gives frequency distributions for the shore development for 95 glacial boreal lakes, and one can note that the mean value is 2.4 and that few lakes have L_d values higher than 8.

The shore development is a useful parameter in contexts of sediment sampling. The purpose of the sampling formula illustrated in fig. 3.20 is to obtain a first estimate of a reasonable number of sediment samples for whole-lake estimates to be used in contexts of aquatic pollution control and monitoring. Thus, the idea is to obtain a general normative reference number and not a thorough description of all factors that may potentially affect the representativity of the sample.

A is the lake area (the formula is mainly meant to be used for lakes larger than 1 km^2); L_d is the shore development. The larger the lake and the more irregular the bottom and also the shoreline, the greater the variability in sediment characteristics and the greater the number of samples needed to obtain a representative mean value for the lake. This relationship is also shown in fig. 3.21 by the regression between the coefficient of variation (CV = SD/MV; SD = the standard deviation; MV = the mean value) for sediment organic content (= loss on ignition) versus, first, in fig. 3.21A, lake area, log(A), and in fig. 3.21B against the shore development (L_d) for nine lakes. From fig. 3.21, one can also note that 55% of the variability among these nine lakes in CV (for the sediment organic content) may be statistically explained by variations among the lakes in lake size (r^2 = 0.55); while 26% of the variability between the CV values can be linked to variability in shore development.

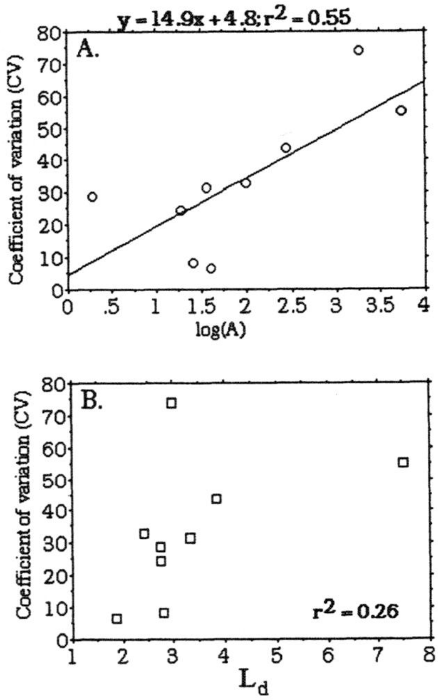

Fig. 3.21. The relationship between the coefficient of variation for within-lake variations in sediment organic content (CV = SD/MV; SD = standard deviation, MV = mean value) versus (upper figure) log(A) and (lower figure) L_d (A = lake area in km^2, L_d = shore development; based on data from nine lakes from Håkanson, 1991).

The morphometric index (MI, dimensionless) gives the relationship between the shoreline length (L_O) and the mean depth (D_{mv}). Several important lake characteristics, such as the total species number of unicellular algae, the bacterial content, the oxygen and temperature gradients show high correlation with the MI value (Shadrin, 1985, 2003).

$$MI = L_O/D_{mv} \quad (3.14)$$

The volume development (V_d, dimensionless) is defined by the ratio between the water volume and the volume of a cone, with a base equal to the water surface area (A in km^2) and with a height equal to the maximum depth (D_{max} in m):

$$V_d = (A \cdot D_{mv} \cdot 0.001)/(A \cdot D_{max} \cdot 0.001 \cdot 1/3) = 3 \cdot D_{mv}/D_{max} \quad (3.15)$$

Where D_{mv} is the mean depth. The volume development describes the form of the lake and is also called the form factor. Table 3.7 and fig. 3.22 give the classification system for lake form based on the relative hypsographic curves and information on how V_d reflects lake form. One can note that lakes with very convex form (VCx) have small V_d values, and vice versa. This relationship is very important, for example, for internal processes. Fig. 3.23 illustrates relative hypsographic curves for four lakes with different forms and hence also V_d values. Lake 1 has a large percentage of the bottom above the wave base, which is the water depth separating areas where wind-induced waves can influence the sediments and above which there are resuspension events. Below the wave base there is a continuous sedimentation of fine materials and no resuspension from surface waves.

Table 3.7. Classification system for defining the form of lake hypsographic curves and corresponding probabilities and based on standard deviations (SD), class limits and V_d values (modified from Håkanson, 1981a).

Lake form	Label	Probability (%)	Class limits (SD)	V_d
Very convex	VCx	6.5	f(-3.0) to f(-1.5)	0.05-0.33
Convex	Cx	24.2	f(-1.5) to f(-0.5)	0.33-0.67
Slightly convex	SCx	38.3	f(-0.5) to f(0.5)	0.67-1.00
Linear	L	24.2	f(0.5) to f(1.5)	1.00-1.33
Concave	C	6.5	f(1.5) to f(3.0)	1.33-2.00

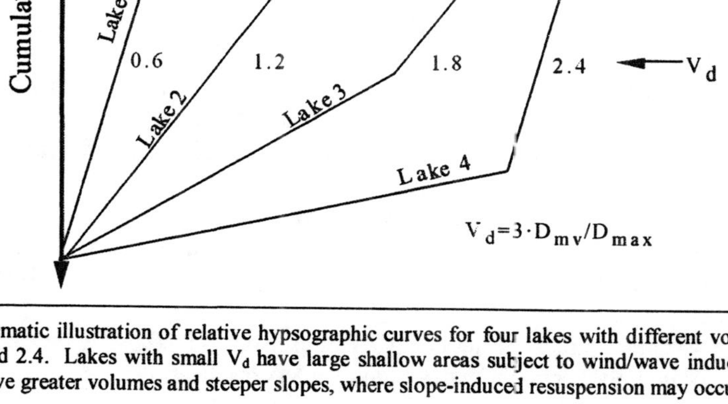

Fig. 3.22. Schematic illustration of relative hypsographic curves for four lakes with different volume development, V_d = 0.6, 1.2, 1.8, and 2.4. Lakes with small V_d have large shallow areas subject to wind/wave induced resuspension. Lakes with high V_d have greater volumes and steeper slopes, where slope-induced resuspension may occur.

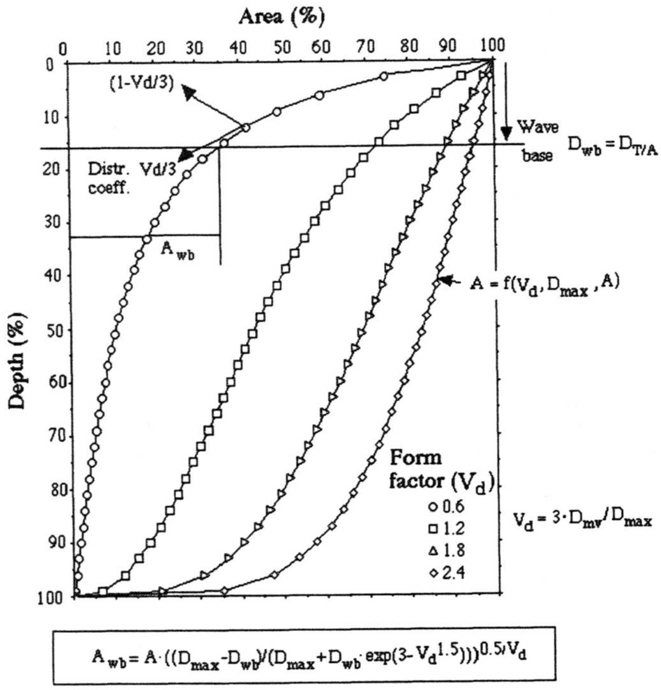

Fig. 3.23. Illustration of how the volume development, V_d, can be used to express the lake form, here given by the relative hypsographic curves (= the depth-area curve) for four lakes. A shallow lake with a small V_d has a relatively large area above the wave base (D_{wb}), where processes of wind/wave-induced resuspension will influence the bottom dynamic conditions, lake Secchi depth and primary and secondary production.

$$A_{wb} = A\cdot((D_{max} - D_{wb})/(D_{max} + D_{wb}\cdot EXP(3 - V_d^{1.5})))^{(0.5/Vd)} \qquad (3.16)$$

The equation expressing the area below the wave base, A_{wb}, as a function of V_d, the area of the lake (A) and the maximum depth of the lake (D_{max}) is given by (see fig. 3.23 and Håkanson, 1999):

Where EXP is the exponent, Deep, U formed lakes generally have smaller areas above the wave base (these are the ET areas, that is, the areas where erosion and transportation processes of fine materials prevail).

Note that the fraction of ET areas also works as a distribution coefficient regulating how much of the settling matter will be deposited either on ET areas above the wave base or below the wave base on the accumulation areas (i.e., areas with a continuous sedimentation of fine materials following Stokes' law), respectively.

The factor $V_d/3$ is also a distribution coefficient regulating the advective fluxes (= resuspension) from ET areas either back to water or to accumulation areas. Eq. 3.16 gives the hypsographic curve as a function of (1) the lake area (A), (2) the maximum depth (D_{max}) and (3) the volume development (V_d).

Fig. 3.24 gives a frequency distribution for V_d values from 95 lakes. One can see that the actual values are approximately normally distributed (the MV/M_{50} ratio is 1.1); the mean V_d value is 1.18, which represents a slightly convex lake form (see table 3.7).

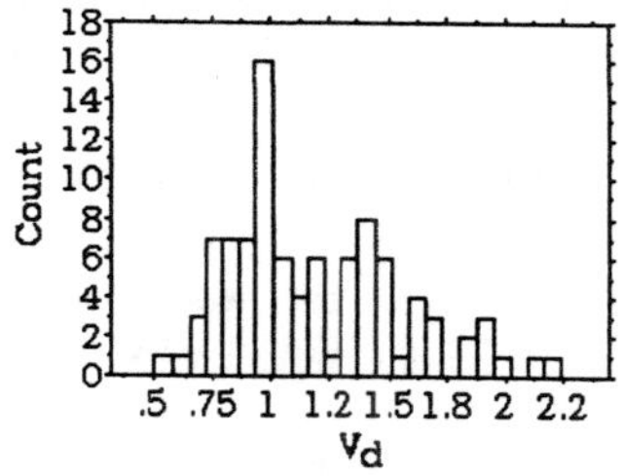

Fig. 3.24. Frequency distribution for the volume development (V_d) based on data from 95 boreal glacial lakes (data from Håkanson and Peters, 1995).

The dynamic ratio (DR) is defined by the ratio between the square-root of the water surface area (in km^2 not in m^2), the $\sqrt{A}$ and the mean depth, D_{mv} (in m; DR = $\sqrt{A}/D_{mv}$). DR is a standard morphometric parameter in contexts of lake resuspension. For lakes > 1 km^2, one can estimate the fraction of the lake bed where resuspension (a key component of internal loading in lakes) is likely to appear (see Håkanson and Jansson, 1983) from:

$$ET = 0.25 \cdot DR \cdot 41^{(0.061/DR)} \quad (3.17)$$

Fig. 3.25 shows that wind/wave-induced ET areas are likely to dominate totally the bottom dynamic conditions in lakes with dynamic ratios higher than 3.8. Slope processes are (see later) known to dominate the bottom dynamic conditions on slopes greater than about 4-5%. From fig. 3.25, one can note that slope-induced ET areas are likely to dominate lakes with DR values lower than 0.052. One can also expect that in all lakes there is a shallow shoreline zone where wind-induced waves will create ET areas, and it is likely that most lakes have at least 15% ET areas. From fig. 3.25, one can also see that if a lake has a DR of 0.25, one can expect that in this lake the ET areas would occupy 15% of the area. If DR is higher or lower than 0.25, the percentage of ET areas is likely to increase.

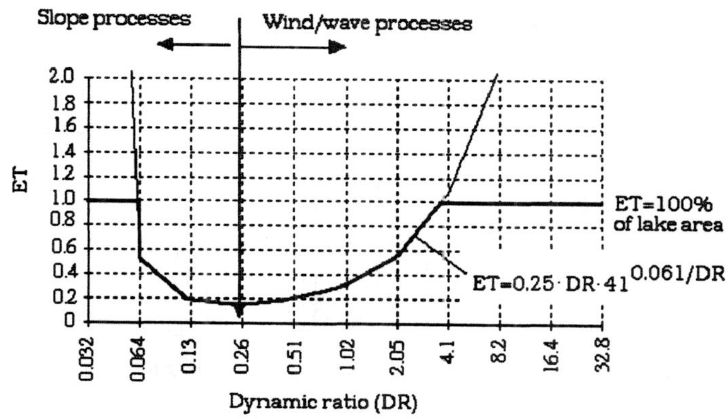

Fig. 3.25. Illustration of the relationship between ET areas and DR. The ET areas are the bottom areas dominated by processes of fine sediment erosion and transportation, and ET is expressed as a fraction of lake area; DR is the dynamic ratio. Resuspension (from wind and wave action) dominates in lakes with high dynamic ratios (DR > 0.25). Turbidity currents related to slope processes are more important in lakes where DR is small. For more information, see Håkanson and Jansson (1983).

Some lakes on earth are extremely large and shallow; and one can assume that Lake Tchad, Africa, with an area of 24,000 km^2, a mean depth of 3.2 m and a dynamic ratio of 44.3, must be one of the most particular cases in this respect. Fig. 3.26 gives a compilation of DR values, lake areas and mean depths from 429 lakes from all over the world (using data from Håkanson and Peters, 1995, and Håkanson and Boulion, 2002). Assuming that this relatively large sample of lakes reflects the population of lakes on earth reasonably well, one can note that it is exceptional for lakes to have DR values larger than 3.8 and smaller than 0.052.

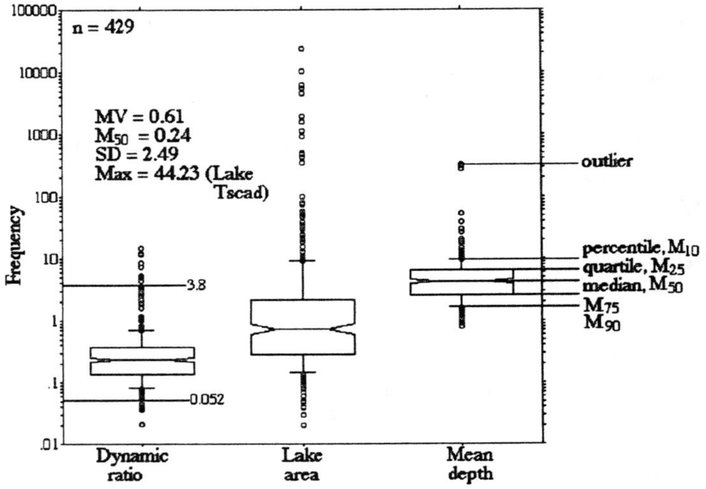

Fig. 3.26. Frequency distribution (box-and-whisker plots) for the dynamic ratio, lake area and mean depth using data from 429 lakes from many parts of the world (data from Håkanson, 2003a).

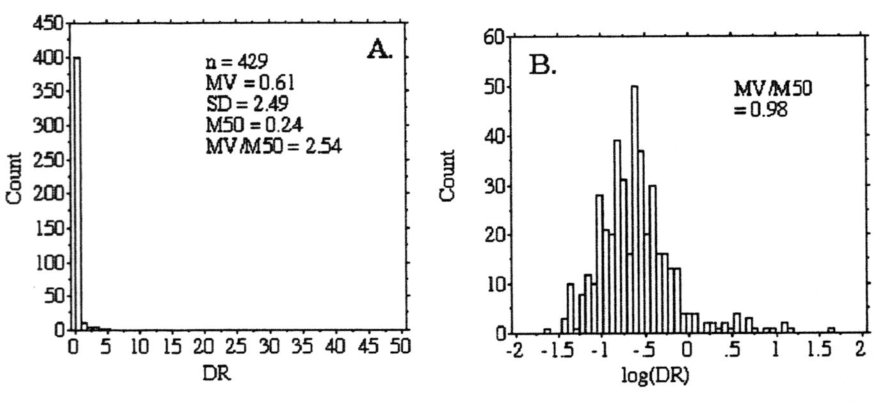

Fig. 3.27. Frequency distributions for the dynamic ratio (DR) based on data from 429 lakes (data from Håkanson, 2003).

From the frequency distribution in fig. 3.27, one can see that the actual DR values are positively skewed (the MV/M_{50} ratio is 2.54; the mean DR value is 0.61).

The wave base (D_{wb} in m) is the depth separating transportation area (with resuspension) from accumulation areas (without resuspension). The wave base has already been discussed (see fig. 3.22). Functionally, this concept is fundamental in limnology since it distinguishes between surface water (epilimnion) and deep-water areas (hypolimnion), see fig. 3.28. To use the thermocline as the key criteria to differentiate between the surface water and the deep water is certainly the most classical approach (Wetzel, 2001). But that does not necessarily imply that it is meaningful and operational in, for example, contexts of mass-balance calculations. This fact is demonstrated in fig. 3.29 using data from Lake Erken, Sweden. For single sites after longer periods without strong winds in the summer, one can certainly see clear thermoclines. However, for many sites over longer periods of time (weeks and months), it is generally very difficult to define the critical water depth requested in mass-balance calculations. Any water depth between 0.1 and 18 m in Lake Erken, which has a maximum depth of 23 m, could be used! The data shown in fig. 3.29 for Lake Erken are included here to emphasize that for this lake, and for lakes in general, the thermocline is a very poor alternative for the requested wave base. Water temperatures do not provide consistent information to define the boundary between surface water and deep water. Most limnologists could probably accept this idea because temperature profiles in most (but maybe not all) lakes look like those shown in fig. 3.29.

The depth of the wave base (D_{wb}) may also be related to concentration profiles. Fig. 3.30 shows vertical variations at six sites on algal biomass (as measured using in vivo fluorescence) in Lake Erken. From this figure, one can draw the same conclusion as from fig. 3.29. That conclusion would also be valid for most substances (not just for algal biomass) in most lakes (not just for Lake Erken). So, for lakes in general, concentration gradients would not provide very useful data to define the requested wave base.

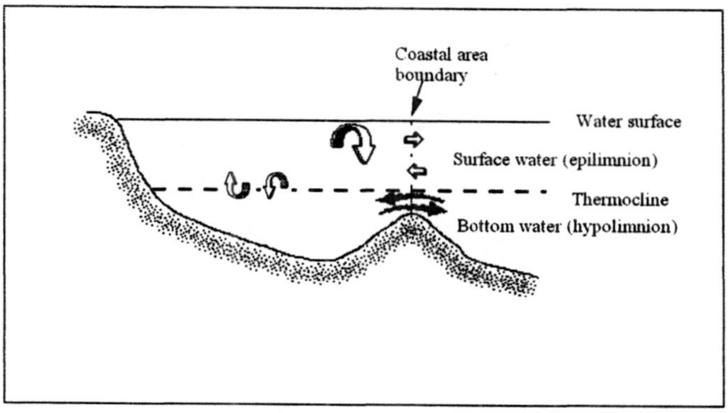

Fig. 3.28. Illustration of important concepts related to stratification, transport and mixing in a coastal area in a large lake: surface water (epilimnion), deep water (hypolimnion) and thermocline.

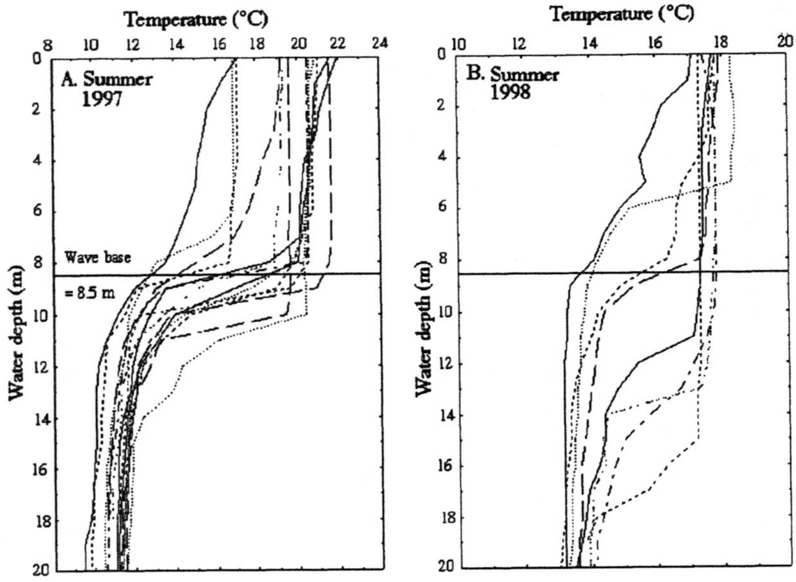

Fig. 3.29. Temperature data from Lake Erken, Sweden (data for the summers of 1997 and 1998 from Håkanson et al., 2003) and illustration of the wave base (D_{wb}) in this lake. Each line represents the temperature profile at the monitoring station at a given sampling occasion.

Fig. 3.30. Data on algal biomass (as determined using in vitro fluorescence) in Lake Erken, Sweden (summer of 2000) from six stations and illustration of the wave base (D_{wb}) in this lake. All the profiles are sampled during the same two-day period in July (data from Håkanson et al., 2003).

The wave base may also be related to the wave equation (see Smith and Sinclair, 1972):

$$g \cdot H/w^2 = 0.0026 \cdot (g \cdot L_{ef}/w^2)^{0.47} \quad (3.18)$$

Where g is the acceleration due to gravity (m/s^2); H is the wave height (in m); w the wind speed (m/s); and L_{ef} the effective fetch (in m; see later).

The wave base (sometimes called the critical water depth) is often set to one-third of the wave length rather than the wave height. In fact, whatever criteria one would use from the wave theory, it would not give a value that could be used in a simple and rational manner in, for instance, mass-balance models based on the ecosystem scale (models valid for entire lakes for longer periods of time, such as weeks and months). Instead, the wave theory gives a whole array of wave heights and wave bases related to different wind situations; and during a period of one week or one month winds can blow from many directions and with many velocities.

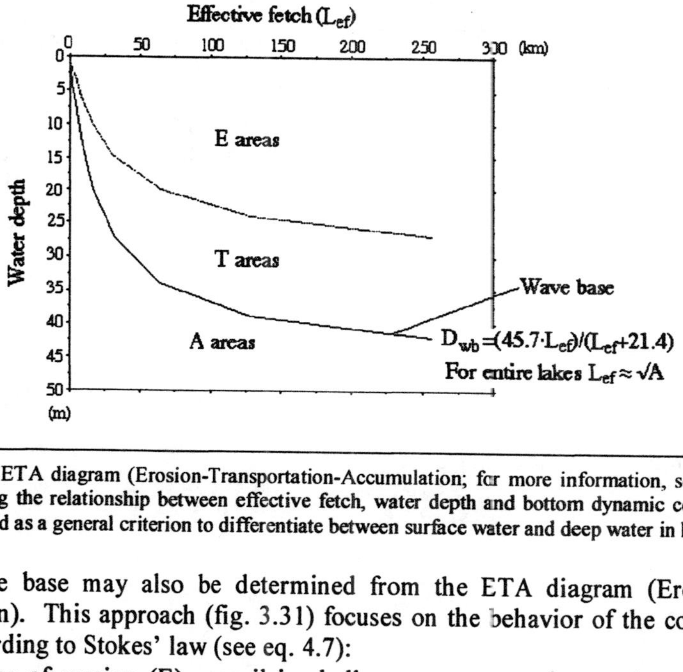

Fig. 3.31. The ETA diagram (Erosion-Transportation-Accumulation; for more information, see Håkanson and Jansson, 1983) illustrating the relationship between effective fetch, water depth and bottom dynamic conditions. The wave base (D_{wb}) can be used as a general criterion to differentiate between surface water and deep water in lakes.

The wave base may also be determined from the ETA diagram (Erosion-Transportation-Accumulation). This approach (fig. 3.31) focuses on the behavior of the cohesive fine materials settling according to Stokes' law (see eq. 4.7):

- Areas of erosion (E) prevail in shallow areas or on slopes where there is no apparent deposition of fine materials but rather a removal of such materials; E areas are generally hard and consist of sand, consolidated clays and/or rocks.
- Areas of transportation (T) prevail where fine materials are deposited periodically (areas of mixed sediments). This bottom type generally dominates where wind/wave action regulates the bottom dynamic conditions. It is sometimes difficult in practice to separate areas of erosion from areas of transportation, and this will not be done here. The water depth separating transportation areas from accumulation areas, the wave base, is a fundamental component in mass-balance calculations.
- Areas of accumulation (A) prevail where the fine materials are deposited continuously (soft bottom areas). It is in these areas (the "end stations") where high concentrations of pollutants are most likely to appear.

The generally hard or sandy sediments within the areas of erosion and transport (ET) often have a low water content, low organic content and low concentrations of nutrients and pollutants (see table 3.8). In connection with a storm, the material on the ET area may be resuspended and transported up and away, generally in the direction towards the accumulation areas in the deeper parts, where continuous deposition occurs. It should also be stressed that fine materials are rarely deposited as a result of simple vertical settling in natural aquatic environments. The horizontal velocity is generally at least 10 times larger, sometimes up to 10,000 times larger, than the vertical component for fine materials or flocs that settle according to Stokes' law (Bloesch and Burns, 1980; Bloesch and Uehlinger, 1986).

Table 3.8. The relationship between bottom dynamic conditions (erosion, transportation and accumulation) and the physical and chemical character of the surficial sediments of Lake Vänern, Sweden. Mean values (MV) and standard deviations (SD) in brackets (data from Håkanson, 1977). n = number of samples.

	Erosion n = 14	**Transportation** n = 16	**Accumulation** n = 84
PHYSICAL PARAMETERS			
Water depth, m	14.9 (9.7)	22.6 (10.3)	42.8 (22.5)
Water content (% ww)	40.3 (14.5)	64.0 (17.9)	84.0 (10.9)
Bulk density (g/cm^3)	1.58 (0.24)	1.27 (0.19)	1.10 (0.08)
Organic content (loss on ignition, % dw)	1.5 (0.8)	4.5 (2.3)	9.7 (5.3)
NUTRIENTS (mg/g dw)			
Nitrogen	0.78 (0.33)	1.14 (0.46)	2.7 (0.53)
CHEMICALLY MOBILE ELEMENTS			
Phosphorus	0.30 (0.50)	1.07 (0.80)	5 (1.0)
METALS (contaminating in this lake; µg/g dw)			
Mercury	0.026 (0.028)	0.27 (0.48)	1.34 (2.13)
Cadmium	<<1 (0.50)	0.7 (0.5)	1.4 (0.8)
Copper	13 (9)	26 (10)	30 (7)
Lead	34 (17)	64 (36)	104 (34)
Zinc	110 (63)	240 (125)	430 (210)
Nickel	22 (9)	31 (12)	28 (7)
METALS (non-contaminating in this lake; µg/g dw)			
Silver	<1	<1	<1
Boron	<20	<20	<20
Beryllium	<1	<1	<1
Bismuth	<10	<10	<10
Chromium	<50	<50	≤50
Molybdenum	<20	<20	<20
Tin	<20	<20	<20
Vanadium	110 (46)	150 (47)	150 (29)

The depth of the wave base, D_{wb}, may be derived from the ETA diagram. D_{wb} (in m) is then calculated from lake area (note that the area should be in km^2 and not in m^2 in eq. 3.19), which is related to the effective fetch (see later) and how winds and waves influences the bottom dynamic conditions:

$$D_{wb} = (45.7 \cdot \sqrt{A})/(21.4+\sqrt{A}) \qquad (3.19)$$

This approach gives one value of the depth of the wave base for any given lake.

Using this criteria to separate the surface water from the deep water, the volume of the surface water may be calculated from (see also fig. 3.23):

$$V_{SW} = V - V_{DW} = A \cdot D_{mv} - A_{wb} \cdot (D_{max} - D_{wb})/3 \qquad (3.20)$$

Where A_{wb} is the area below the wave base (the accumulation areas) calculated from eq. 3.16. To calculate A_{wb} (and hence also the area above the wave base, $A_{ET} = A - A_{wb}$), one needs data on the lake area (A), the maximum depth (D_{max}), the wave base (D_{wb}) and the form factor V_d.

The following processes influencing internal loading may be accounted for in an approach to predict the ET areas in lakes (see Håkanson and Jansson, 1983):

1. an energy factor related to the effective fetch and the wave base (see the ETA diagram in fig. 3.31),
2. a form factor related to the percentage of the lake bed above the wave base (see fig. 3.23), and
3. a lake slope factor related to the fact that slope-induced transportation (turbidity currents) may appear on bottoms inclining more than 4-5% (fig. 3.32).

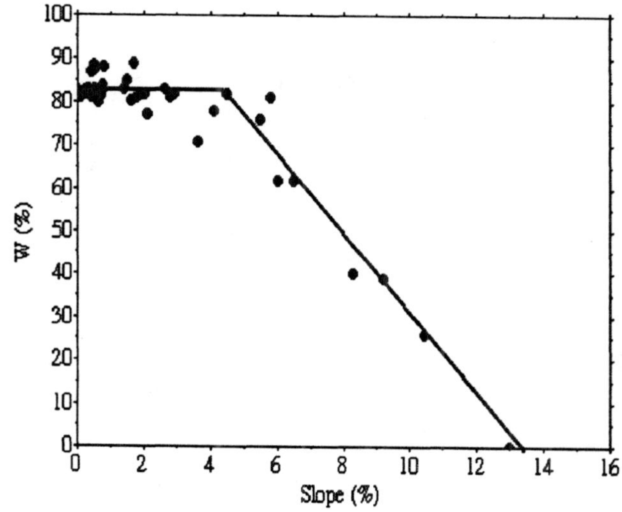

Fig. 3.32. The relationship between slope and water content of surficial sediments (0-1 cm) within a test area in central Lake Vänern, Sweden (data from Håkanson and Jansson, 1983). The position of this test area is such that one would expect uniform sedimentological conditions; the major factor controlling the physical sediment character (here given by water content) is the slope.

For lakes larger than 1 km^2, eq. 3.17 based on the dynamic ratio (fig. 3.25) has been tested with good results to calculate ET. Another approach may also be used, the one utilizing the wave base (from fig. 3.31) and the form of the lake, as illustrated in fig. 3.23. Then, the depth of the wave base (D_{wb}) is set equal to the depth separating transportation areas from accumulation areas as given by eq. 3.19. An evident boundary condition for this approach to calculate the ET areas is that if $D_{wb} > D_{max}$, then $D_{wb} = D_{max}$.

There are also two boundary conditions for ET (= the fraction of ET areas in the lake):

If $ET > 0.99$ then $ET = 0.99$
If $ET < 0.15$ then $ET = 0.15$

From fig. 3.25, one can conclude that ET areas are generally larger than 15% ($ET = 0.15$) of the lake area since there is always a shore zone dominated by wind/wave activities. For practical and functional reasons, one can generally also find sheltered areas, macrophyte beds and deep holes with more or less continuous sedimentation, that is, areas which actually function as A areas, so the upper boundary limit for ET may be set at $ET = 0.99$ rather than at $ET = 1$.

As mentioned, the value for the ET areas is used as a distribution coefficient in mass-balance models. It regulates the sedimentation of the particulate fraction of the substance X either to accumulation areas or to ET areas. These approaches to calculate ET are summarized in fig. 3.33.

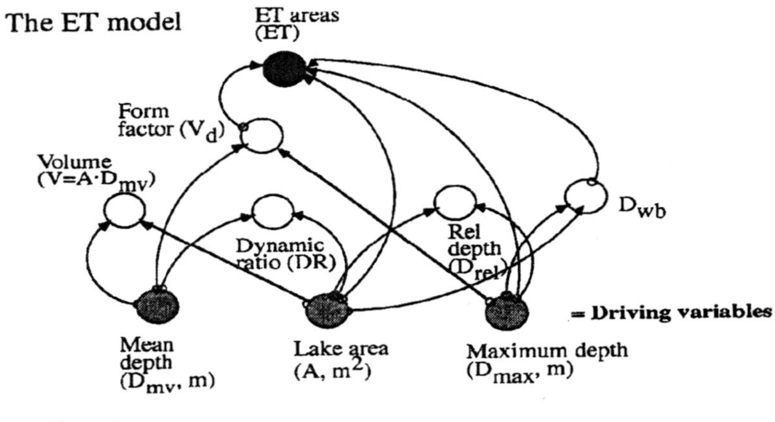

Equations:

$DR = \sqrt{(A \cdot 10^6)}/D_{mv}$
$D_{rel} = (D_{max} \cdot \sqrt{\pi})/(20 \cdot \sqrt{(A \cdot 10^6)})$
$D_{wb} = (45.7 \cdot \sqrt{(A \cdot 10^6)}/(21.4+\sqrt{(A \cdot 10^6)})$; if $D_{wb} > D_{max}$ then D_{max}
$V_d = 3 \cdot D_{mv}/D_{max}$
If $A \geq 1$ km^2 then $ET = 0.25 \cdot DR \cdot 41^{(0.061/DR)}$
If $A < 1$ km^2 then
$ET= 1-(A \cdot ((D_{max}-D_{wb})/(D_{max}+D_{wb} \cdot EXP(3-Vd^{1..5})))^{(0.5/Vd)})/A$
If $ET > 0.99$ then $ET = 0.99$
If $ET < 0.15$ then $ET = 0.15$

Fig. 3.33. Compilation of the two approaches to calculate ET, either based on the dynamic ratio (DR) for lakes larger than 1 km^2 and/or based on the depth of the wave base (D_{wb}) and the relative hypsographic curve (for more information, see Håkanson and Boulion, 2002).

The depth of the wave base (D_{wb}) may also be estimated from sediment samples collected at various sites at different water depths. As indicated in fig. 3.34, the water content (and all physical sediment variables related to the water content, such as the organic content, the bulk density and the grain size) of surface sediments (0-1 cm) is also closely linked to the morphometry. The sediment water content is low at small water depths, as in the ET areas. At the wave base (D_{wb}, 27 m for Lake Vänern as shown in fig. 3.34), the water content is generally about 10% lower than in the deepest part of the lake where the characteristic water content, W_k (= 89.0% in Lake Vänern) is attained. So, the depth of wave base may be estimated from sediment samples as the depth where the water content is 10% lower than the W_k value (89-10 = 79% in Lake Vänern).

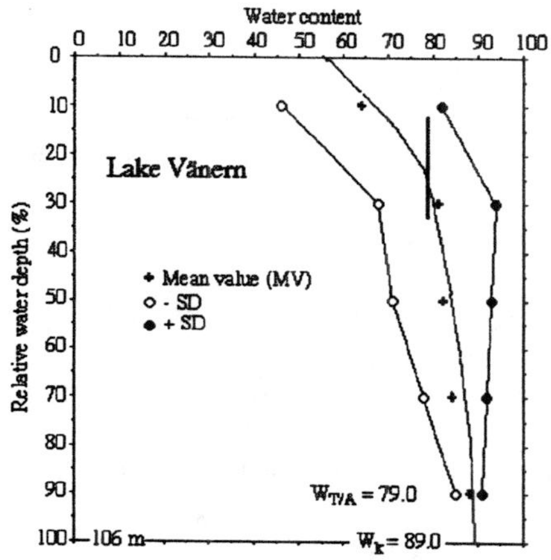

Fig. 3.34. The water content of surface sediments (0-1 cm) as a function of relative water depth in Lake Vänern. The characteristic water content, W_k, is the asymptotic value of the water content when the relative water depth approaches 100% (after Håkanson, 1981b).

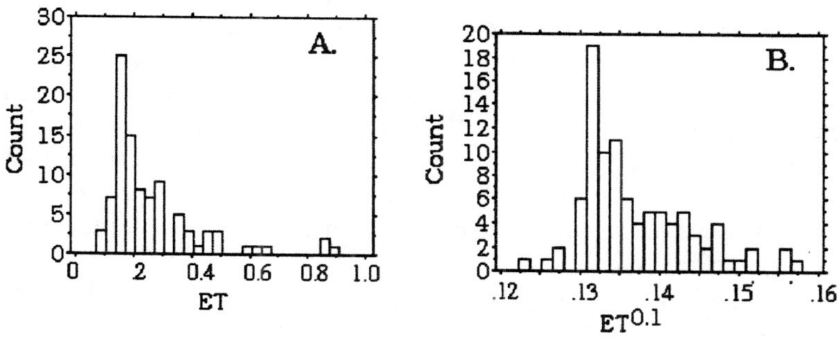

Fig. 3.35. Frequency distribution for (A) the fraction of ET areas (i.e., areas where processes of fine sediment erosion and transportation prevail) and (B) for $ET^{0.1}$ based on data from 95 boreal glacial lakes (data from Håkanson and Peters, 1995).

Fig. 3.35 gives two frequency distributions for ET based on data from 95 lakes. One can note that the mean value is 0.26 and that the transformation $ET^{0.1}$ gives a frequency distribution which is close to normal (MV/M_{50} = 1.01).

The effective fetch (L_{ef} in km) is often defined according to a method introduced by the Beach Erosion Board (1972; see fig. 3.36). The effective fetch gives a more representative measure of how winds govern waves (wave length, wave height, etc.) than the effective length (L_{el}), since several wind directions are taken into account. Using traditional methods, it is relatively easy to estimate the effective fetch by means of a special transparent paper shown in fig. 3.37. The central radial of this transparent paper is put in the main wind direction or, if the maximum effective fetch is requested, in the direction which gives the highest L_{ef} value. Then the distance (x in km) from the given station to land (or to islands) is measured for every deviation angle a_i, where a_i is ± 6, 12, 18, 24, 30, 36 and 42 degrees. L_{ef} may then be calculated from:

$$L_{ef} = \Sigma x_i \cos(a_i) / (\Sigma \cos(a_i)) \cdot SC' \quad (3.21)$$

Where

$\Sigma\cos(a_i)$	= 13.5, a calculation constant.
SC'	= the scale constant; if the calculations are done on a map in scale 1:250,000,
then SC'	= 2.5.

The areal distribution of the maximum effective fetch (L_{xef}) in Lake Vänern is shown in fig. 3.38. One can note that L_{xef} attains the highest values close to the shoreline and the minimum values in the central part of a basin. This relationship is important in, for example, contexts of shore erosion and morphology (see fig. 3.39), for bottom dynamic conditions (erosion-transportation-accumulation; see fig. 3.40), and hence also for internal processes, mass-balance calculations, sediment sampling and sediment pollution.

For entire lakes, the mean effective fetch may be estimated by $\sqrt{A}$ (see fig. 3.31). In a round lake, the requested value should be somewhat lower than the diameter ($d = 2 \cdot r$; r = the radius); the area A is $\pi \cdot r^2$ and hence $d = 1.13 \cdot \sqrt{A}$ and the mean lake fetch approximately $\sqrt{A}$.

Table 3.8 gives a compilation of physical sediment variables, water content, bulk density and organic content, in areas of erosion, transportation and accumulation in Lake Vänern and the water depths of these bottom dynamic zones. The table also gives corresponding data on nutrients (nitrogen and phosphorus). In addition, it shows that phosphorus is a very mobile element in lake sediments, reflecting predominant redox conditions rather than the depositional patterns. Concentrations of less mobile, contaminating metals and non-contaminating metals in the sediments of this lake are also given in table 3.8 for the three bottom dynamic zones.

The bottom dynamic map for Lake Vänern (fig. 3.40) demonstrates very clearly that the erosion areas prevail in shallow and exposed positions, whereas the accumulation areas dominate at greater water depths. To stress the close link between morphometry, bottom dynamic conditions and the areal variations in sediment pollution, fig. 3.41 gives a map on that matter for mercury, the main pollutant in this lake.

One can note the low concentrations in erosion-area sediments and that the highest mercury concentrations appear close to the polluting sites in the northern part of the lake and in areas of accumulation in sheltered coastal areas or at larger water depths in the open water area. This distribution indicates the great importance of the next morphometric variable, the filter factor, which expresses how an archipelago area would act as an energy filter between the open water areas and a more enclosed coastal area.

Sample site: 100

a_i	$\cos(a_i)$	x_i	$x_i \cdot \cos(a_i)$
42	0.743	2.3	1.7
36	0.809	2.9	2.3
30	0.866	3.4	2.9
24	0.914	4.5	4.1
18	0.951	23.9	22.7
12	0.978	21.9	20.4
6	0.995	20.9	20.8
0	1.000	21.0	21.0
6	0.995	19.0	18.9
12	0.978	18.4	18.0
18	0.951	18.4	17.5
24	0.914	12.0	11.0
30	0.866	10.3	8.9
36	0.809	6.7	5.4
42	0.743	7.7	5.7
Sum: 13.512			181.5

$L_{ef} = 2.5 \cdot 181.5/13.5$

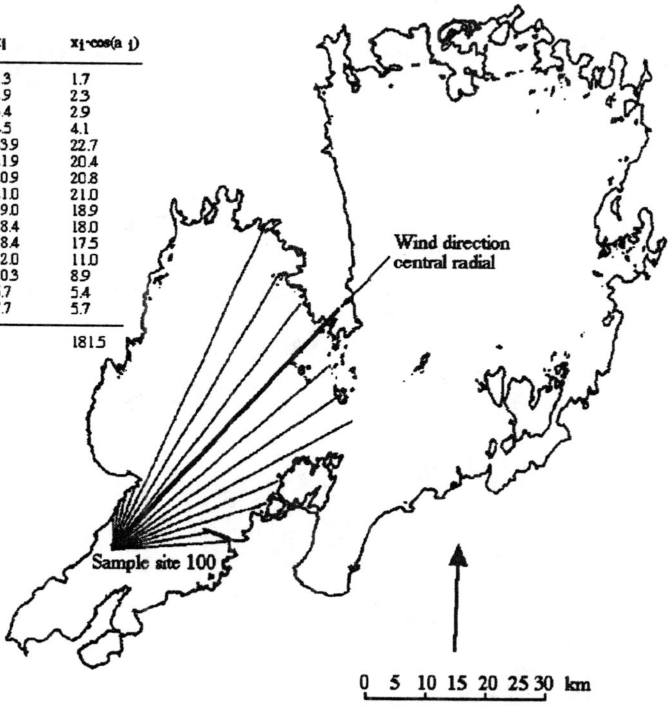

Fig. 3.36. Determination of the effective fetch (L_{ef}) at site 100 in Lake Vänern.

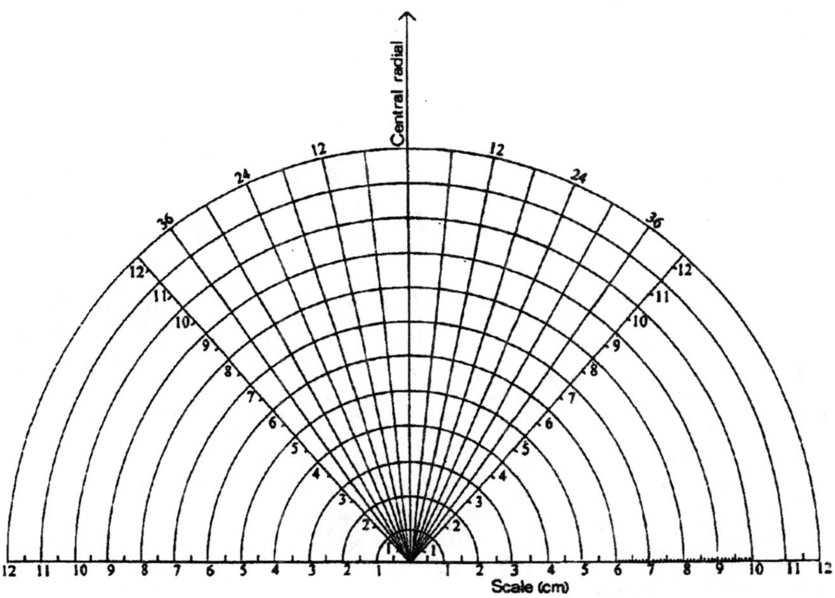

Fig. 3.37. Diagram illustrating a simple technique to determine the effective fetch (L_{ef}). This diagram should be on a transparent paper and put on the map.

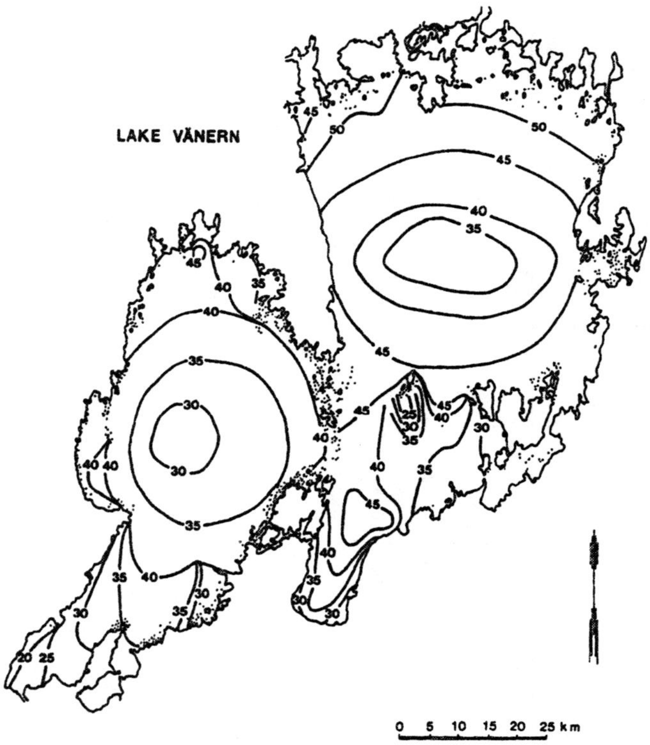

Fig. 3.38. The areal distribution of the effective fetch (L_{ef}) in Lake Vänern.

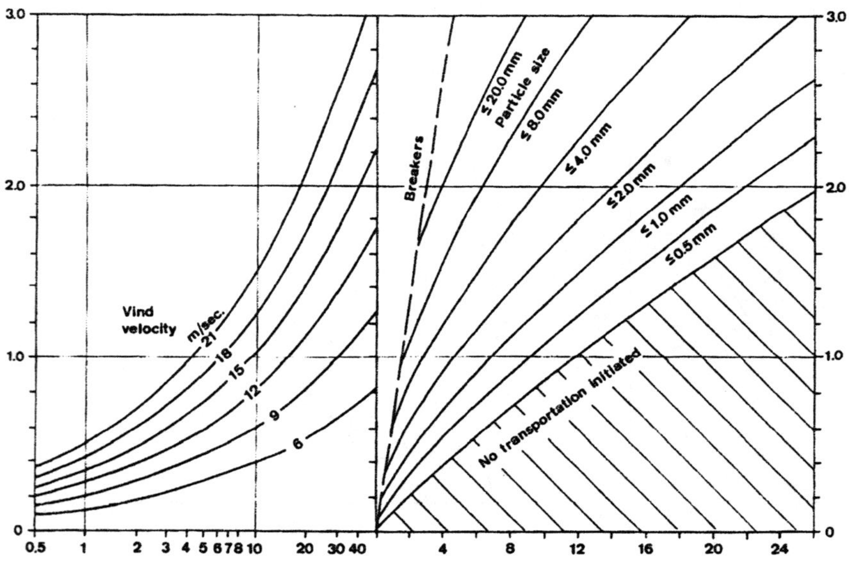

Fig. 3.39. The relationship between (left) the effective fetch (Lef) and the wave height and (right) between the wave height and the bottom dynamics in the shore zone (after Norrman, 1964 and Håkanson, 1981a).

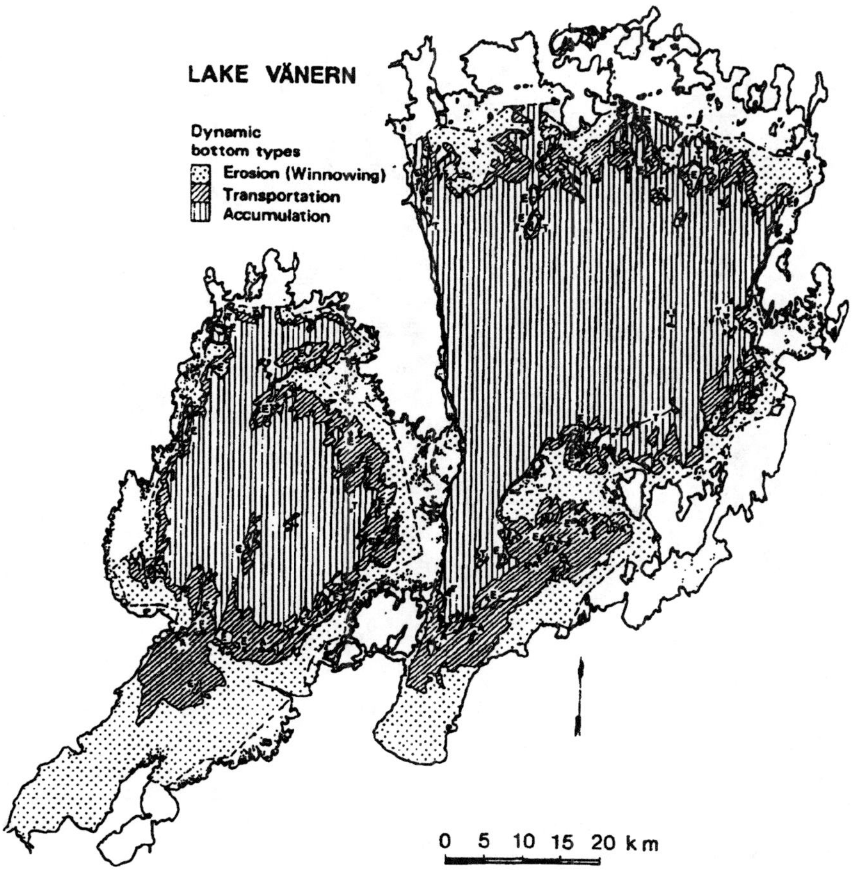

Fig. 3.40. The areal distribution of bottom dynamic types (i.e., areas where fine sediment erosion, transport and accumulation prevail) in Lake Vänern.

The filter factor (FF, km^3) is a measure related to both the effective fetch and the opening towards the open water areas. It may be used to quantify the amount of wave energy from the open water areas that reaches a coastal area in a large lake after passage through the surrounding archipelago. The effective fetch gives the wind/wave impact at one single site. The filter factor, on the other hand, accounts for the wind/wave impact on an entire coastal area. The filter factor is defined as:

$$FF = \sum(\sum(\cos(a_i)\cdot x_i\,)\cdot A_{sa})_j \qquad (3.22)$$

Where a_i are angles in radians and x_i the distances to land in km for a imaginary group of radials starting from the mid-point of one of the opening sections of the coastal area (A_{sa}) (see fig. 3.42). The index i ranges from 1 to 315, which means that the angle varies between $-\pi/2$ and $\pi/2$ radians in relation to the normal of the opening section. The sum of $\cos(a_i)\cdot x_i$ for the 315 angles is multiplied by the section area of the opening (A_{sa}). This product is calculated for all the openings in the coastal area (j), and the sum of them is the filter factor.

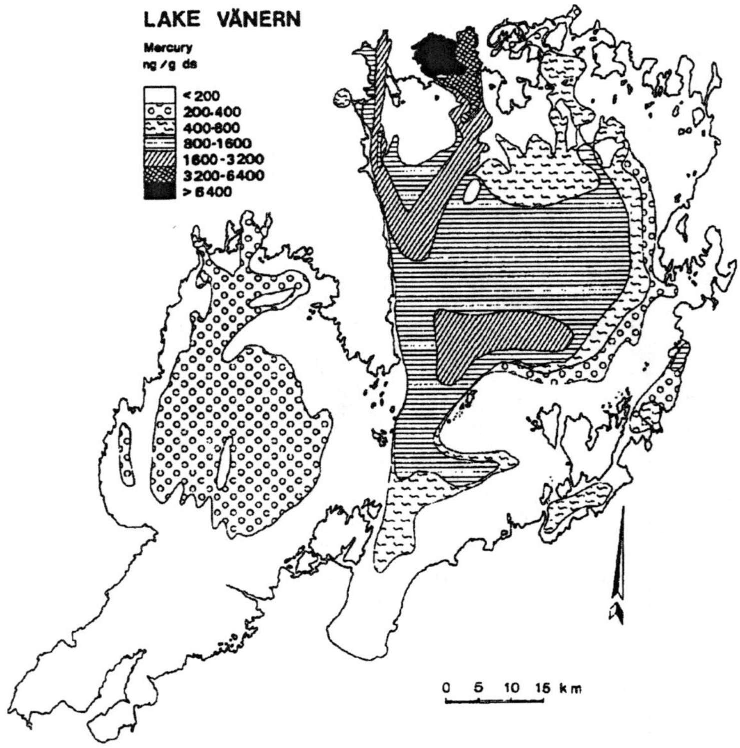

Fig. 3.41. The areal distribution of mercury in the surficial sediments (0-1 cm) in Lake Vänern. The dominant mercury source is a chlor-alkali industry in the north (from Håkanson, 1977).

A large section area and an open archipelago give a high FF value. It should be noted that the filter factor cannot be calculated if one or more radials are endless. However, the influence on the waves and the bottom dynamic conditions is more or less constant if the fetch exceeds 300 km (see fig. 3.31); so the maximum length of a radial may be set to 300 km in the calculation of the filter factor using eq. 3.22.

Lake bottom roughness (Ro, dimensionless) measures the degree of irregularity of the lake bottom in analogy with the shore development, which gives the irregularity of the shoreline. From fig. 3.43, one can note that there is also a very strong relationship between these two measures of topographical irregularity (the more irregular the bottom topography, the more irregular the shoreline). The Ro value for entire lakes is defined by:

$$Ro = (0.165 \cdot (L_{ci} + 2) \cdot L_{ctot}) / (D_{50} \cdot \sqrt{A}) \qquad (3.23)$$

Where

L_{ci} = the countourline interval in m;
L_{ctot} = the total normalized length of all contour lines including the shoreline in km;
D_{50} = the median depth in m;
A = the lake water surface area in km^2.

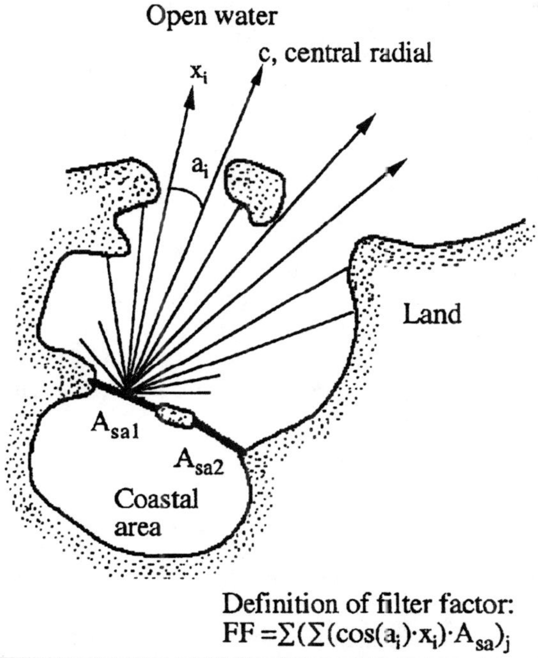

Fig. 3.42. Illustration of the filter factor (FF) determined for one of the two section areas in this coastal area; i = the number of radials at each opening; j = the number of openings; j = 2 in this example.

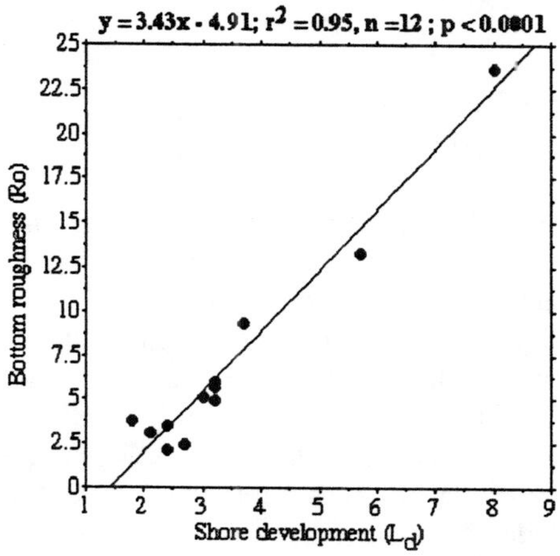

Fig. 3.43. The relationship (regression line, r^2, n and p) between the shore development (L_d) and the lake bottom roughness (Ro; data from Håkanson, 1974).

Table 3.9. The normalized lake bottom roughness, Ro, for eight Swedish lakes.

Lake	Lake area (km^2)	Ro-value
Vänern	5648	16.0
Vättern	1856	7.1
Hjälmaren	84	12.0
Roxen	94.7	6.9
Ekoln	18.6	2.7
Saxen	7.1	7.3
Velen	2.8	8.3
Munksjön	1.1	4.7

Using traditional methods, the total length of all contour lines (excluding the shoreline) may be estimated from:

$$L_{ctot}' = 2.1 \cdot L_d^{1.5} \cdot \sum_{i=1}^{n} \sqrt{A_{ci}} \qquad (3.24)$$

Where L_d is the shore development and A_i is the area defined by contour line i.

To obtain a Ro value to be used for comparative purposes, it is necessary to utilize a constant number of contour lines for all lakes. The rule of thumb is that n should be 5.

The lake bottom roughness of Lake Vänern is comparatively large, Ro = 16. This value is high compared to other lakes (see table 3.9).

The Ro value should not be used to compare the degree of bottom irregularity of various sub-basins in a lake. If this information is required, the form roughness (Ro_f) is a preferable measure.

$$Ro_f = (0.165 \cdot (L_{ci}+2) \cdot L_{ctot})/A_i \qquad (3.25)$$

Where

L_{ci} = the contour line interval in m;
L_{ctot} = the total normalized length of all contour lines including the shoreline in km;
A_i = the lake water surface area of the given sub-basin in km^2.

The Ro_f value may provide important information in contexts of sediment sampling. More samples are needed in areas with high Ro_f to determine representative mean values. Fig. 3.44 gives the areal distribution of the form roughness in Lake Vänern. This map has been constructed in scale 1:60,000 using squares of 21.44 km^2. The greatest depth difference (dD) within each square has been determined; 5 contour lines have been drawn in each square as the positive integer of dD/5; the total length of the contour lines has been determined in each square; the Ro_f values have been determined; and the isopleths for the following Ro_f values have been drawn, 8, 16, 24 and 32. Fig. 3.44 shows the relative differences in bottom irregularity within the system. The obtained Ro_f values depend on the choice of the base area. The median Ro_f value in lake Vänern is 23.

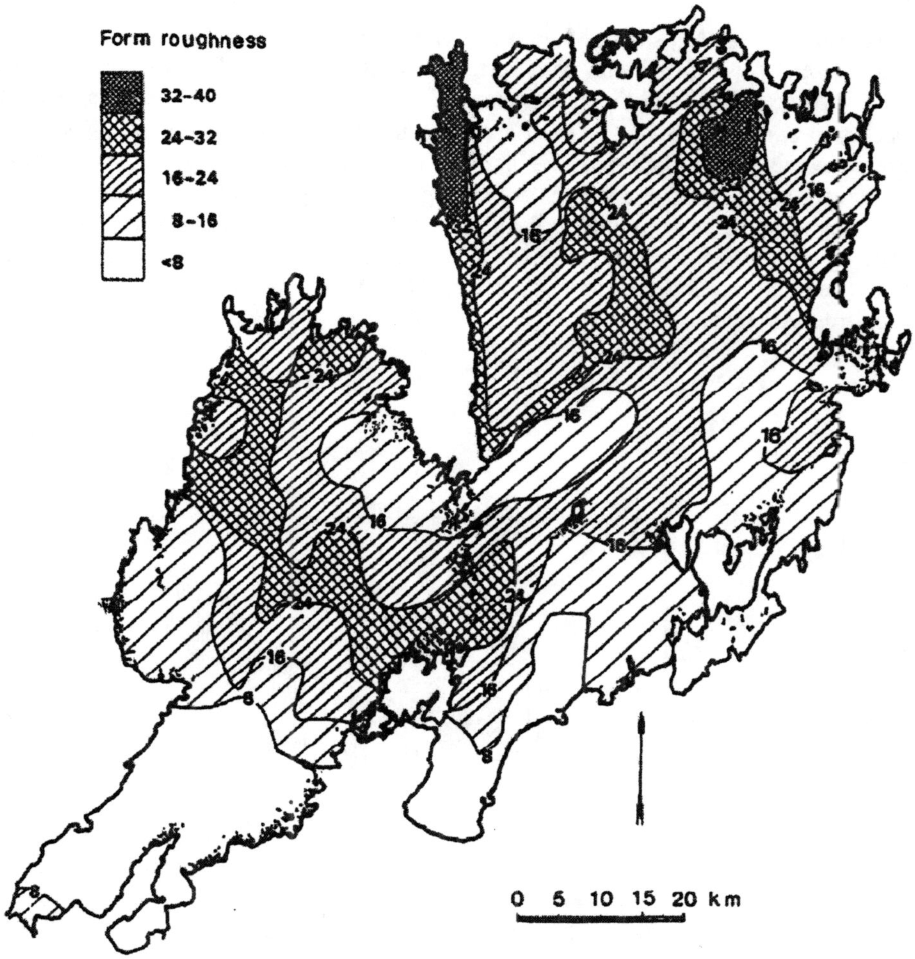

Fig. 3.44. The areal distribution of the form roughness (Ro_f) in Lake Vänern.

The lake bottom roughness may also be estimated directly from echograms accordingly (from Galvenius, 1975).

$$Ro' = L_x/L_{sx} + Ly/L_{sy} - 1 \quad (3.26)$$

Where

Ro' = the lake bottom roughness;
Lx = the total length of the actual bottom profile along the x-direction in km;
L_y = the total length of the actual bottom profile along the y-direction in km;
L_{sx} = the total straight-line length of the echosounded tracks along the x-direction in km;
L_{sx} = the total straight-line length of the echosounded tracks along the y-direction in km.

3.1.3.2. Parameters related to measures of area and volume

The section area (A_{sa}; see fig. 3.45 for illustration; generally in m^2 or km^2) may be calculated by planimeter, or using GIS with both vector data and raster data for any given opening between the defined coastal area and the open water area. If there are several openings, A_{sa} is the sum of all section areas ($A_{sa} = \Sigma A_{sai}$). The coordinates for the openings to the surrounding open water areas, stored as vector data, may be used to calculate the opening lengths (according to the Pythagorean theorem). These lengths are then multiplied by the corresponding mean depths along the openings (estimated according to the principle described below) and summed.

The exposure (Ex) of a coastal area is related to the total section area (A_{sa}) and defined as the ratio between the section area and either the 3D bottom area (A_{3D}) (see eq. 3.8) or the water-surface area (A) from:

$$Ex = A_{sa}/A \qquad (3.27)$$

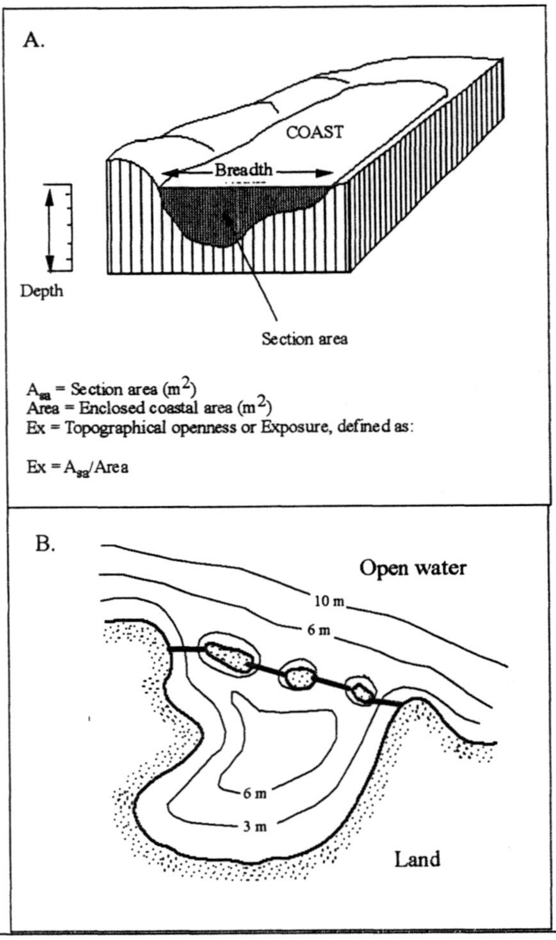

Fig. 3.45. A. Illustration of the section area as the boundary between the enclosed coastal area and the open water area and definition of the exposure (Ex). B. A coastal area with four openings and four section areas.

Large, enclosed and deep coastal areas have low Ex values while smaller, more exposed and shallower coastal areas have higher Ex values. The exposure is very important for the water exchange and the bottom dynamic conditions in coastal areas (see Håkanson, 2000). It is sometimes expressed in percent ($Ex = 100 \cdot A_{sa}/A$).

The hysographic curve (= the depth-area curve) is constructed by putting the depth on the negative ordinate (y-axis) and the cumulative area on the positive abscissa (x-axis). The hypsographic curve illustrates form elements of the basin and provides a means whereby the area of any depth level may be determined. The hypsographic curve may also be used in determinations of lake volume. The hypsographic curve for Lake Vänern is illustrated in fig. 3.46A.

The percentage hypsographic curve is obtained by putting the cumulative area in percentage on the positive x-axis and the depth on the negative y-axis. The percentage hypsographic curve for Lake Vänern is given in fig. 3.46B.

The relative hypsographic curve is obtained by putting depth as well as the cumulative area in percentage on the two axes. The relative hypsographic curve illustrates lake form and nothing but lake form. It is the base for the definition of the lake form concept already discussed (see tables 3.3 and 3.4). The relative hypsographic curve for Lake Vänern is shown in fig. 3.46C.

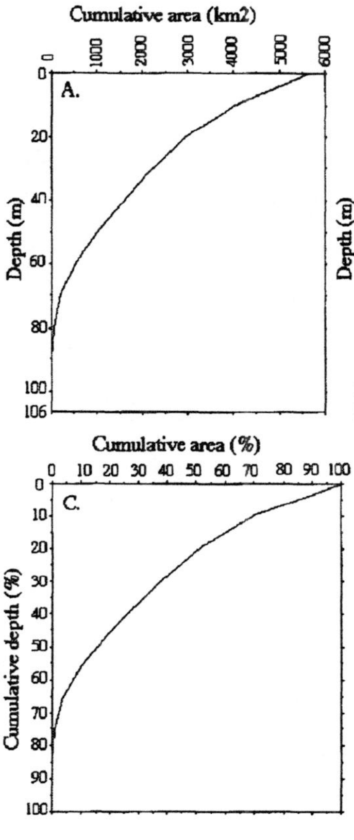

Fig. 3.46. Depth-area curves for Lake Vänern.

The mean lake form, f(MV), signifies that there is a 50% chance for an unknown lake to have a relative hypsographic curve above (on the convex side) or below (on the concave side) of the f(MV) curve. The statistical deviation forms corresponding to ±0.5, ±1.0, ±1.5, ±2.0 and ±3.0 standard deviations (SD) are labeled f(±0.5), f(±1.0), etc. Fig. 3.47 gives a schematical bathymetric interpretation of three different lake forms. A lake of the f(-3.0) type has one or more areally limited deep holes, but is generally very shallow. A lake of the f(3.0) type, on the other hand, is trough-like with steep inclining wall and a very plane areally dominating bottom. Terminology and class limits for the various lake form have been given in table 3.7 and are illustrated in fig. 3.48. Very convex lakes, VCx, have relative hypsographic curves between the limitation lines f(-3.0) and f(-1.5). The probability that an unknown lake shall be of this VCx type is only 6.5%. Most lakes are of the slightly convex (SCx) type, 38.5%. Thus, the mean lake form, f(MV), is slightly convex.

The relative hypsographic curves for ten large lakes are shown in fig. 3.49. As already discussed, the lake form concept is of fundamental importance for lake function, since the form influences many important lake processes, such as resuspension, the Secchi depth, the concentration of suspended particulate matter, internal loading of nutrients, primary and secondary production and hence biomasses of key functional organisms. This relationship will be discussed in greater detail in chapters 4 and 5.

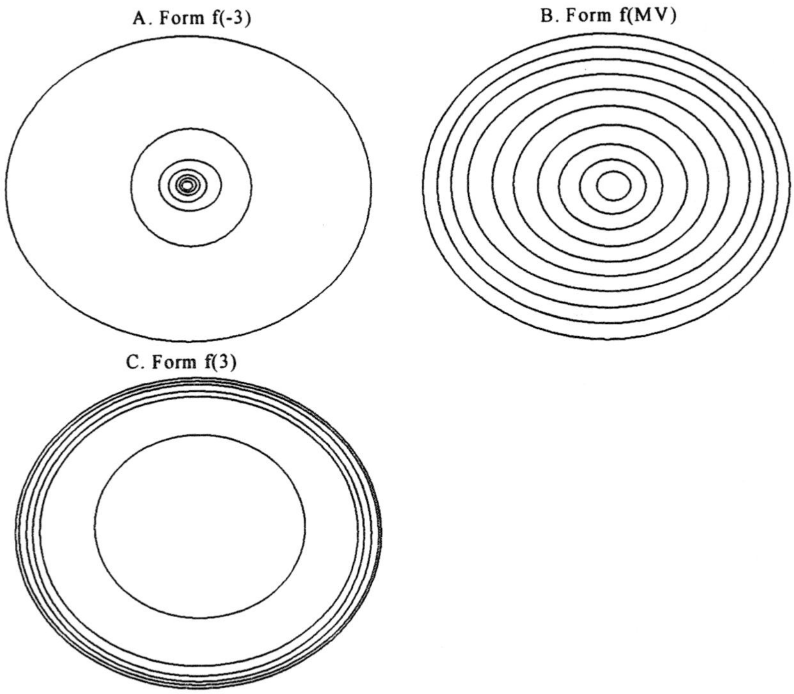

Fig. 3.47. Schematical bathymetric interpretation of three statistical lake forms.

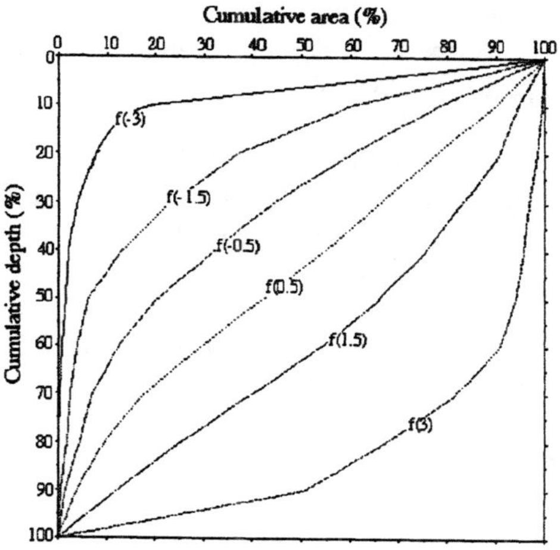

Fig. 3.48. Terminology and class limits for the classification system for lake forms based on relative hypsographic curves.

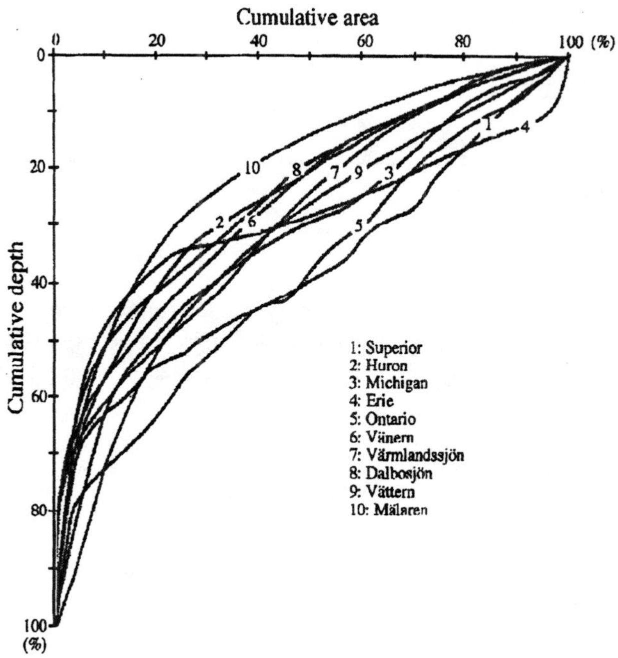

Fig. 3.49. Relative hypsographic curves for ten large lakes.

The volume curve illustrates the relationship between depth and volume in the same way as the hypsographic curve shows the relationship between depth and area. The cumulative volume at each depth interval is given on the positive x-axis and the depth on the negative y-axis. The volume curve for Lake Vänern is illustrated in fig. 3.50.

The percentage volume curve is given by putting the cumulative volume in percentage on the positive x-axis and the depth on the negative y-axis (see fig. 3.50B). The relative volume curve is obtained by putting cumulative volume as well as cumulative depth in percentages on the two axes (see fig. 3.50A).

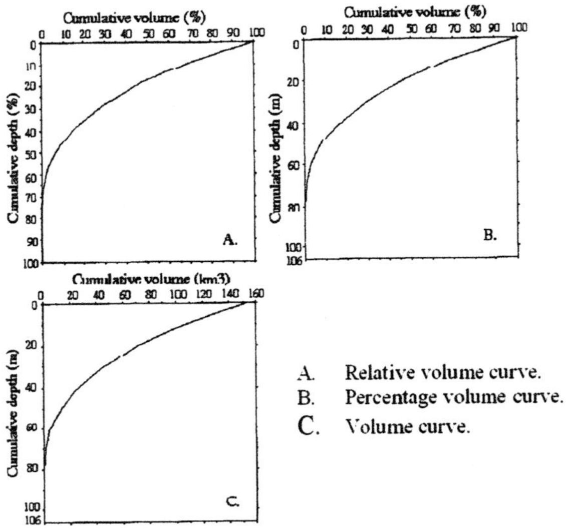

Fig. 3.50. Volume curves for Lake Vänern.

3.1.3.3. Slope conditions

The slope (S_i) for an arbitrary site in a lake may be determined directly from the echogram provided that the track follows the same direction as the major slope axis, which can be identified from the bathymetric map. The slope may be given in degrees or as a percentage of the height versus the length (see fig. 3.51).

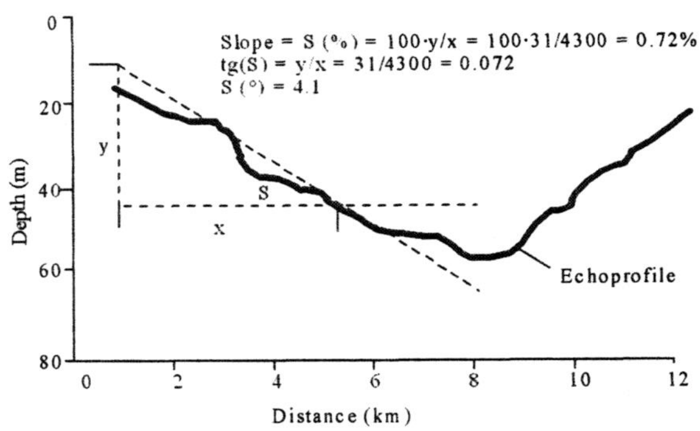

Fig. 3.51. Illustration of the determination of the slope from an echogram.

The slope (S in %, S_p) between two contour lines (of length L_1 and L_2 in km) in a map can be determined from the following formula (see fig. 3.52):

$$S_p = L_{ci} \cdot (L_1 + L_2) / (20 \cdot A_{12}) \qquad (3.28)$$

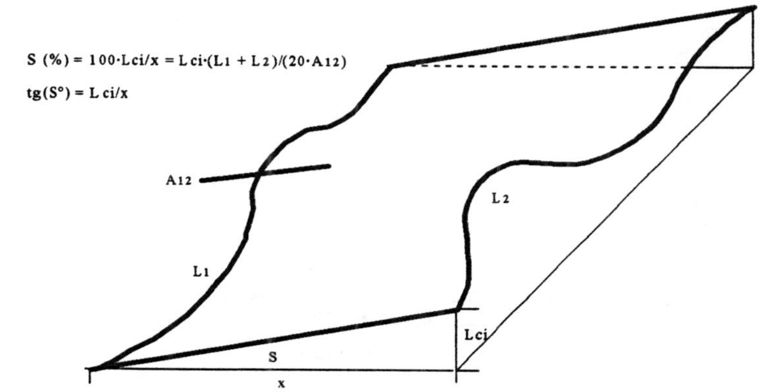

Fig. 3.52. Illustration of determination of the slope between two contour lines.

Where L_{ci} is the contour line interval (m) and A_{12} is the area between the two contour lines in km^2.

The slope is an important morphometric parameter in contexts of lake sedimentology (see also Gilbert, 1975; Sturm, 1975). Fig. 3.32 exemplifies this fact and gives the relationship between the slope and the physical status of surficial sediments within a test area in central Lake Vänern. The position of this test area is such that one would expect uniform sedimentological conditions (accumulation area) had it not been for the slope. The major factor controlling the physical character of these sediments (water content, bulk density, grain size, etc.) is the slope. No recent fine materials will stay permanently on these slopes if the inclination is higher than 4-5%.

The <u>mean slope</u> (S_{mv} in %) may be defined for an entire lake (or coastal area) by:

$$S_{mv} = (L_0 + 2 \cdot L_{ctot}) \cdot D_{max} / (20 \cdot n \cdot A) \qquad (3.29)$$

Where

- L_0 = the normalized shoreline length in km;
- L_{ctot} = the total normalized length of all contour lines excluding the shoreline in km;
- D_{max} = the maximum depth in m;
- n = the number of contour lines;
- A = the lake water surface area in km^2.

Using GIS, the mean slope may be estimated by vector operations on data from the raster file (see Pilesjö et al., 1991).

The <u>median slope</u> (S_{50}) can be determined from the <u>slope curve</u> (see fig. 3.53). The median slope signifies that 50% of the lake area inclines more than S_{50}. The median slope is 0.75% in Lake Vänern.

The mean slope and the median slope are primarily descriptive lake parameters.

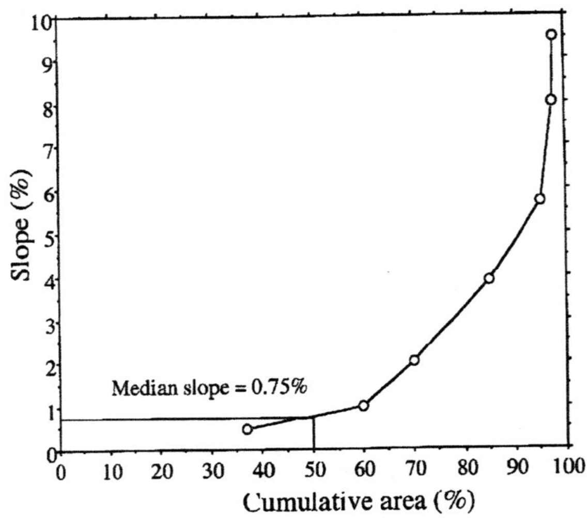

Fig. 3.53. Determination of the median slope from the slope curve for Lake Vänern.

3.1.3.4. Islands

The insulosity (Ins in %), the percentage of the total lake area (A_{tot}) that is occupied by islands, islets and rocks (A_{itot}; see below for definitions), is estimated by dividing the area of all islands by the total area.

$$Ins = 100 \cdot A_{itot}/A_{tot} \quad (3.30)$$

The following classification may be used to distinguish islands, islets and rocks (from Håkanson, 1974):

Area (km^2)	Name
< 0.001	Rocks
0.01-0.01.1	Islets
0.01-1	Islands
>1	Large islands

This nomenclature has an advantage over, for example, the one proposed by Hedenstierna (1948) because it is based on a specified area division, which simplifies statistical calculations. It also agrees with the prevalent linguistic usage, at least as well as any other classifications system.

The mean area of islands (A_{ismv} in km^2) is estimated by dividing the area of all islands (A_{itot}) by the number of islands (ni) in the lake:

$$A_{ismv} = A_{itot}/n_i \quad (3.31)$$

3.2. Compilation and conclusion

Table 3.10 gives comparative results when five standard morphometric parameters have been determined using manual methods and GIS methods. The correspondence is generally good with r^2 values higher than 0.88.

Table 3.10. Differences between manual and computer-based estimations of five standard morphometric parameters in 20 coastal areas (from Pilesjö et al., 1991).

Parameter	r^2
Maximum depth (m)	0.94
Mean depth (m)	0.92
Total area (km^2)	0.90
Water surface area (km^2)	0.90
Water volume (km^3)	0.88

Fig. 3.54 gives a compilation of key morphometrical parameters related to coastal areas in large lakes. One can note that many of the morphometrical parameters (area, volume, mean depth, maximum depth, shore development, etc.) can be used both for entire lakes and for coastal areas. There are also variables that apply only for coastal areas, such as the filter factor and the section area. There are models to predict the theoretical surface water retention time, the deep water retention time and the bottom dynamic conditions for coastal areas based on readily determined coast-morphometric parameters (see fig. 2.5).

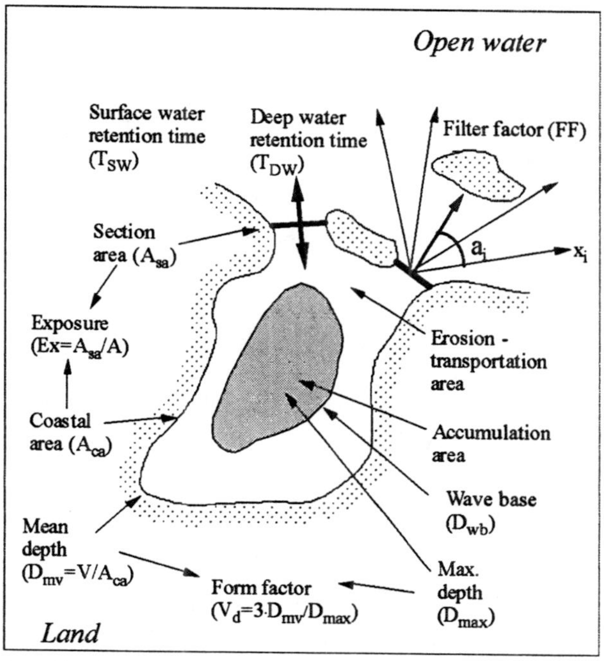

Fig. 3.54. Illustration of key morphometrical parameters, which can be determined either manually or from digitized bathymetric map using GIS methods. An important criterion in contexts of coastal area studies, it is used to define the boundary line, i.e., where the coastal area ends and the open water areas begin. This approach uses the "topographical bottleneck" method to do this whereby the limitation lines are drawn where the exposure ($Ex = A_{sa}/A$) attains the minimum value.

This chapter has demonstrated that it is both possible and meaningful to obtain accurate estimations of morphometric parameters using both traditional methods and GIS methods. The calculated "length parameters" using GIS seem to be reasonably scale independent. There is no doubt that a computer-based geographical information system for lake and coastal studies is more time- and cost-effective than traditional, manual methods.

4. The Importance Of Lake Morphometry In Limnology 1 – Ranking Based On Statistical Analyses

The aim of this chapter is to use empirical data and statistical analyses to demonstrate how lake morphometric parameters influence variables expressing important lake characteristics. Many factors (x-variables) could potentially influence the variability among and within lakes of a given target y-variable. The statistical analysis based on empirical data can be used to rank the importance of how the x-variables influence y. In these contexts, one must clearly differentiate between a statistical analysis and a causal analysis. Statistical analyses, like regressions, can never mechanistically "explain" why certain x-variables end up with a high correlation towards y; but results from correlations and regressions can provide important information for further mechanistic interpretations, understanding and modelling. Such applications will be demonstrated in this chapter. The next chapter will focus more on the mechanistic aspects of how morphometry influences lake processes. It is assumed that the reader has some basic knowledge in statistical analyses, but appendix 8.1 is included to give a short recapitulation of basic concepts in regression analysis.

This chapter will also address effect-load-sensitivity (ELS) analyses since lake morphometric data are often essential regulators between load and effect, i.e., they are key sensitivity factors. The same load of a contaminant may cause different effects in lakes of different size and form.

Section 4.3 will discuss an interpretational key to explain how different factors (such as parameters expressing lake morphometry, catchment areas characteristics and climatological variables) cause variations for given limnological variables (y). Section 4.3 also addresses variations within lakes among lakes, as well as the basic elements of mass-balance calculations in an attempt to clarify how lake morphometry can be expected to influence transport processes in lakes relative to other factors.

Normally, one cannot get information on seasonal variability from map parameters quantifying lake form and catchment area characteristics, such as percentage of lakes, mires, forests and soil types. Map parameters are generally used to explain variations among lakes. There is, however, a technique to estimate seasonal changes in water discharge and lake temperature (both epilimnetic and hypolimnetic temperatures) from map parameters (such as altitude, latitude, mean annual precipitation and continentality) combined with special (calibrated) calculation routines. Since this technique mainly does not concern lake morphometric parameters (but altitude, latitude, etc.), it is not given in the main text. More information is given, instead, in appendix 8.2.

In statistical analyses, it is often important to identify clusters or "families" of variables showing large internal co-variations. This process will be discussed in a section 4.3, and it is also part of appendix 8.1.

4.1. Introduction – on predictive power and practical usefulness of models

A central problem in communications among scientists, and a key reason for much misunderstanding, has to do with scale. This problem is evident in practically all matters dealing with, for example, phytoplankton, bacterioplankton and zooplankton interactions. Very many species, groups and size classes are involved, and differently in different systems. To sort this out in great detail at the scale dealing with hourly or daily changes for individual species is a very difficult task indeed for one system; and to try to do this in a general, predictive manner for many systems is even more difficult. The regressions discussed in this chapter do not address the finer details of trophic interactions or the changes from one sample site to the next, or from one day to the next, or from one species to the next, in a given well-studied lake. This problem is further motivated in fig. 4.1.

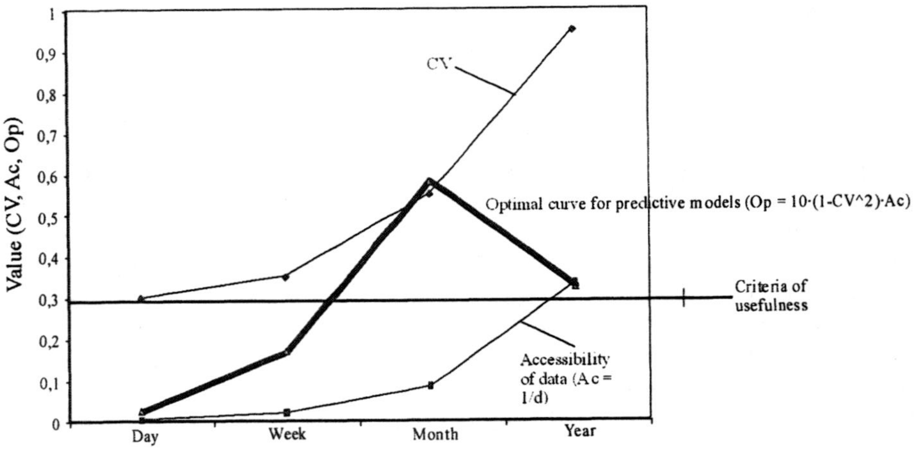

Fig. 4.1. Illustration of factors regulating the optimal size of practically useful predictive models in ecology and radioecology.

This figure illustrates some important general concepts related to the "philosophy" behind the concept of scale in lake studies. The curve marked CV in fig. 4.1 illustrates how the uncertainty in empirical data increases if the lake data are collected during a day (a relatively small CV of 0.3; CV = the coefficient of variation; CV = SD/MV; SD = standard deviation; MV = mean value), during weekly sampling within a lake (a higher CV, 0.35), for monthly sampling (a still higher value, CV = 0.55) and for samples collected over a year (CV = 0.95). The curve called "accessibility" illustrates that the overall sampling effort would increase, and the money and work related to the accessibility of data needed to run and test a model would also increase, for models based on daily predictions (using series of measured climatological data) relative to models for long-term predictions based on fewer and more readily available data from standard maps and/or monitoring programs. The boldface curve is meant to illustrate one of many possible ways to define the optimal size of a predictive model accounting for these two factors, the uncertainty in the empirical data (as given by CV) and the accessibility of the data (as given by Ac = 1/d, where d is simply the number of days used for sampling; note that it is not meaningful to determine a mean value or a standard deviation from a sampling with fewer than three samples). The horizontal line illustrates that any given model developer or user would have a set of criteria to define the practical usefulness of the model; different users may have different criteria. Given the curve for practical usefulness in this figure, one can note that models striving for short-term (hourly - daily) predictions are not very useful due to too high demands related to data accessibility, and that models based on annual predictions would be sub-optimal due to the high uncertainties in the empirical data.

4.2. Effect-load-sensitivity analyses and the role of lake morphometry

Differential equations are often used to quantify fluxes (g X /yr), amounts (g X) and concentrations (g X/m^3) of all types of materials (such as toxins and nutrients), but not generally ecosystem effect variables. Regressions based on empirical data are often necessary to relate concentrations of chemicals in lake water or sediments to target effect variables. In theory, both these model approaches (see fig. 4.2A and B) may be used for the effect-load-sensitivity analyses (ELS) provided that at least one operationally defined ecological effect variable relevant for the load variables(s) in question is included in the model. Step B in fig. 4.2 illustrates a regression, and this figure can also serve as an illustration to the main focus of this chapter.

Ideally, the effect variable should express the production or biomass of defined functional organisms (preferably fish at the top trophic level, see fig. 4.2C), which characterize a given lake. Such effect variables will be discussed in chapter 5 in relation to aspects of lake size and form. Fig. 4.2D illustrates schematically what will be shown in greater detail later in this chapter and also in chapter 5, that two lakes of different morphometry are likely to react differently to a change in the load of phosphorus to the lake. The classical approach to carry out environmental consequence analyses is, however, to use dynamic mass-balance models to predict concentrations of pollutants and empirical models (like regressions) to link these concentrations to measured data on the effect variables. In contexts of water management, one can also for practical, economic or scientific reasons seek simpler but relevant operational effect variables, like mean concentrations of a given toxic substance (or substances) in predatory fish, which will be discussed later in this chapter

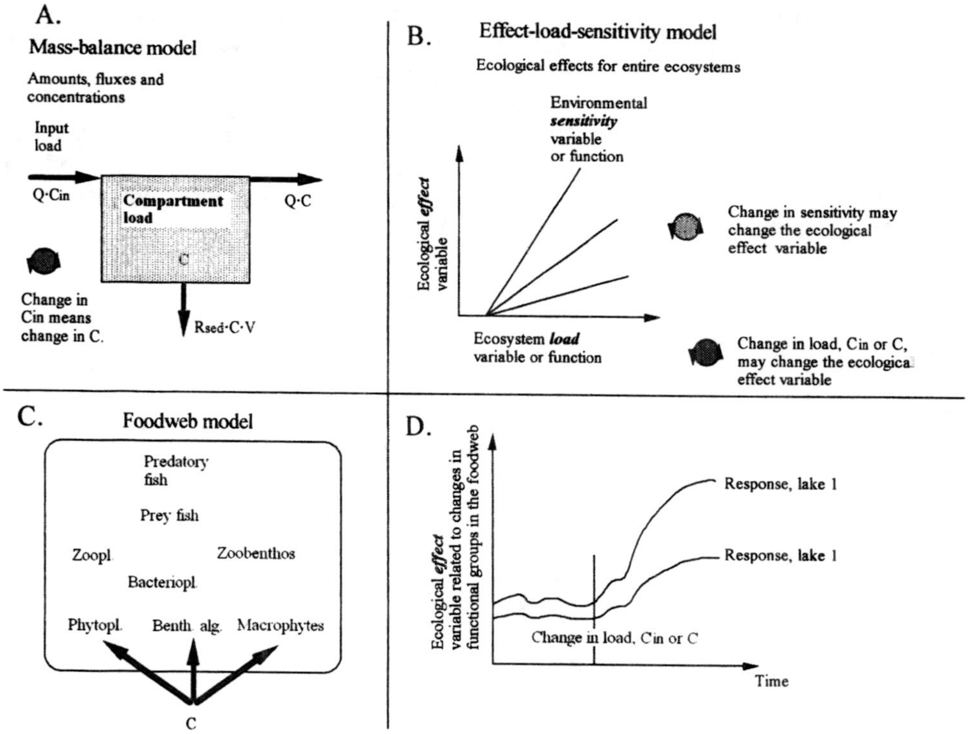

Fig. 4.2. Illustration of the fundamental difference between dynamic mass-balance models (fig. A) and effect-load-sensitivity models (ELS) based on regressions (B) and ELS models related to dynamic foodweb models (C) and (D) how changes in the load at a given time may cause different responses in the lake foodweb in lakes with different morphometry (lake 1 compared to lake 2). The wheels indicate that by means of remedial measures one may reduce the load variable in dynamic models and the load and the sensitivity variables in ELS models.

It should be stressed that ELS analyses are of fundamental importance in lake management and that ELS models are essential tools to examine consequences of remedial measures that may influence a target effect variable. One can reduce negative ecosystem effects of toxic substances by reducing the load to the lake or by changing the sensitivity (for example, by adding lime or nutrients to the lake). One cannot generally change the lake morphometry, but lakes of different size and form will react differently to remedial measures and it is essential to know this so that one can have realistic expectations of the remedial measures for a given lake.

Richard Vollenweider presented the first load models for phosphorus for lakes in the late 1960s (Vollenweider, 1968). By means of simple mass-balance calculations and statistical regressions, he could demonstrate that in many lakes, eutrophication could be reversed by reducing the input of total phosphorus (TP) to the lakes so that the mean lake annual concentration of TP could be lowered. Since then, many studies have demonstrated where the Vollenweider approach can and cannot be used. Thus, total phosphorus is recognized as the most crucial limiting nutrient for lake primary production (Schindler, 1977, 1978; Bierman, 1980; Chapra, 1980; Boynton et al., 1982; Boers et al., 1993). The famous Vollenweider model (and later versions, such as OECD, 1982), and the analysis behind this load model, constitutes a fundamental base for practically all environmental assessments of phosphorus in lakes. This approach has played a paramount role in lake management. The interesting part, however, is not to predict a concentration of a chemical element like total phosphorus, but to predict ecological effects related to lake TP. From table 4.1, it is clear that lake TP can be quantitatively linked to almost any ecological effect variable or key functional group characterizing lake ecosystems. It is evident that the TP concentration can be influenced by emissions from many types of sources, like point sources (domestic sewage, industries and fish farms), atmospheric deposition (to the lake surface and the catchment area), internal loading (linked to resuspension, diffusion, etc.) and, often most important, tributary input. The characteristics of the catchment, like bedrocks, soils, land use, etc., regulate the concentration of TP in the tributaries to the lake.

Table 4.1. Selected regressions based on extensive empirical data illustrating the key role of phosphorus in lake studies. Many biological variables whose determinations normally require extensive and expensive field and laboratory work may be estimated or predicted from one key abiotic state variable, lake characteristic TP (in μg/l = mg/m^3). Some variables may be predicted with great precision (high r^2), others with much less. OECD82 is OECD (1982), Vor77 is Vorobev (1977) and H&B02 is Håkanson and Boulion (2002). PrimP = primary production (in g ww/m^2·yr), Maccov = Macrophyte cover of lake bed (in % lake area), SecMV = mean annual Secchi depth (in m), D_{mv} = mean lake depth (in m), Col = lake color (mg Pt/l), DC = distribution coefficient (dimensionless), n = number of lakes used in the regression, ww = wet weight and dw = dry weight.

y-value	Equation	Range for x	r^2	n	Units	Reference
Equations based on phosphorus used as norms in LakeWeb:						
Chlorophyll (summer mean)	$=0.28\cdot TP^{0.96}$	2.5-100	0.77	77	mg ww/m^3	OECD82
Chlorophyll (summer max.)	$=0.64\cdot TP^{1.05}$	2.5-100	0.81	50	mg ww/m^3	OECD82
Chlorophyll (weekly mean)	$=0.5\cdot TP$ [(0.64+0.28)/2 ≈ 0.5; (0.96+1.05)/2 ≈ 1)]				mg ww/m^3	H&B02
Max. prim. prod. (TP>10)	$=20\cdot TP-71$	7-200	0.95	38	mg C/m^3·d	Peters (1986)
Max. prim. prod. (TP<10)	$=0.85\cdot TP^{1.4}$				mg C/m^3·d	Peters (1986)
Mean prim. prod. (TP>10)	$=10\cdot TP-79$	7-200	0.94	38	m C/m^3·d	Peters (1986)
Mean prim. prod. (TP<10)	$=0.85\cdot TP^{1.4}$				mg C/m^3·d	Peters (1986)
Phytoplankton	$=30\cdot TP^{1.4}$	3-80	0.88	27	mg ww/m^3	Peters (1986)
Phytoplankton	$=30\cdot TP^{(1.4-0.1\cdot(TP/80-1))}$				mg ww/m^3	H&B02
Bacterioplankton (Col < 50)	$=0.90\cdot TP^{0.66}$	3-100	0.83	12	mill./ml	Peters (1986)
Bacterioplankton	$=(1+0.25\cdot(Col/50-1)))\cdot 0.90\cdot TP^{0.66}$(if color > 50				mg Pt/l)	H&B02
Zooplankton, herbivores	$=0.77\cdot 38\cdot TP^{0.64}$	3-80	0.86	12	mg ww/m^3	Peters (1986)
Zooplankton, predators	$=0.23\cdot 38\cdot TP^{0.64}$ (the distr. coeff. is 0.77)				mg ww/m^3	H&B02
Zoobenthos	$=810\cdot TP^{0.71}$	3-100	0.48	38	mg ww/m^2	Peters (1986)
Zoobenthos	$=(1-0.5\cdot(TP/100-1))\cdot 810\cdot TP^{0.71}$ (if TP >100)				mg ww/m^2	H&B02
Fish	$=590\cdot TP^{0.71}$	10-550	0.75	18	mg ww/m^2	Peters (1986)
Fish yield	$=7.1\cdot TP$	8-550	0.87	21	mg ww/m^2·yr	Peters (1986)
Other equations used as norms in LakeWeb:						
Fish yield	$=0.0023\cdot PrimP^{0.9}$	170-14000	0.64	66	mg ww/m^2·yr	H&B02
Prey fish	=DC·fish biomass (default DC=0.73)				mg ww/m^2	H&B02
Predatory fish	=(1-DC)·fish biomass				mg ww/m^2	H&B02
Macrophyte cover	$= 0.50\cdot(Sec_{MV}/D_{mv})$			229	%	Vor77
Macrophytes	$=1.37\cdot \log(Mac_{cov})+3.58$				g ww/m^2·yr	H&B02
Zooplankton, herbivores	$=0.15\cdot(PrimP\cdot 1000)^{0.86}$	13-15000	0.61	42	g ww/m^2·yr	H&B02
Zooplankton, predators	$=0.076\cdot(PrimP\cdot 1000)^{0.84}$	2-3000	0.43	42	g ww/m^2·yr	H&B02
Other equations:						
Nanoplankton	$=17\cdot TP^{1.3}$	3-80	0.93	23	mg ww/m^3	Peters (1986)
Macrozooplankton	$=20\cdot TP^{0.65}$	3-80	0.86	12	mg ww/m^3	Peters (1986)
Net plankton	$=8.6\cdot TP^{1.7}$	3-80	0.82	23	mg ww/m^3	Peters (1986)
Crustacean plankton	$=5.7\cdot TP^{0.91}$	3-300	0.72	49	mg ww/m^3	Peters (1986)
Microzooplankton	$=17\cdot TP^{0.71}$	3-80	0.72	12	mg ww/m^3	Peters (1986)
Blue-greens	$=43\cdot TP^{0.98}$	8-1300	0.71	29	mg ww/m^3	Peters (1986)

Fig. 4.3 gives the principles of an ELS model illustrated as a ELS diagram. Environmental goals (generally set by National Environmental Protection Agencies) should concern the ecological effect variables and not the load variables, since one and the same load may cause different effects in ecosystems of different sensitivities. The figure also gives some of the ELS variables that may be used for metals in lakes, and it stresses the role of morphometry. From this diagram, important concepts like natural background concentration and critical load can be scientifically defined. There are, to the best of the author's knowledge, no practically useful validated ELS models for metals today (except for mercury and radiocesium, which will be discussed later on). This lack means that there exists ample room for speculations about cause and effect, and about the best strategies to remediate aquatic systems polluted with metals.

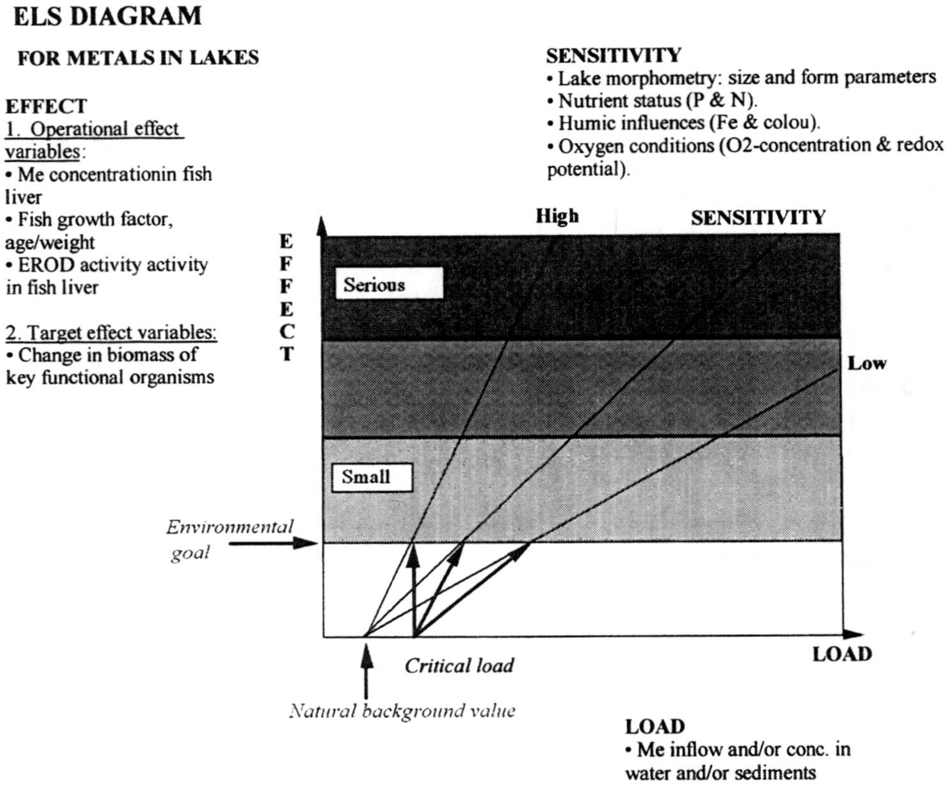

Fig. 4.3. The basic set-up of an ELS diagram from an ELS model. This figure also gives examples of effect, load and sensitivity variables that may be used for ELS models for metals in lakes and stresses the role of lake morphometry in contexts of sensitivity (= vulnerability) to regulate the "road between" metal load and effects of metals in lakes.

In contexts of ELS analyses, the primary interest is not on site-specific conditions ("the sampling bottle"), on the individual, organ or cell level, but at the ecosystem level. That perspective should be of main interest from a management point of view where questions are posed concerning the status of larger water bodies (ecosystems), and the remedial actions that could be used in practice to improve the conditions in such systems.

The predictive accuracy of the model should always be stated. The predictive power is generally determined by means of statistical tests, where model predictions of the effect variable are compared to empirical data of the effect variable. Models that provide r^2 values higher than about 0.75 (and p values lower than 0.05; Prairie, 1996) could generally be used in practice in management for predictions in individual ecosystems. Models providing lower r^2 values (but still p values < 0.05) could be used for regional predictions, when predictive failure in individual ecosystems could be accepted (Håkanson and Peters, 1995). It is generally not possible to derive ELS models which apply with equal success to all types of ecosystems. Therefore, the operational range, the domain, of the model must be explicitly given to avoid abuse of the model for ecosystems for which it was never intended to be used. If dynamic (time-dependent) mass-balance models can meet these requirements, they would generally be preferable to statistical/empirical models because they can provide better understanding of mechanisms and processes. Chapter 5 will focus on the role of lake morphometry in dynamic models, and this chapter examines morphometry in empirical models.

4.3. An interpretational key to the causes of variations in lake variables

This section will give an interpretational key, or template, to explain causes for variations among and within lakes in important lake variables (y).

A relevant question in predictive ecosystem modelling is: How well can a given target variable y be predicted? It is evident that many factors have to be considered, such as sampling (such as the number of samples), analysis (the precision in determining y), model structure (how and which model variables x_i are included), the reliability of the model variables and the statistical methods used to define predictive success. In this section, the r^2 value (from regressions where modelled values are compared to empirical data) is used as a standard criterion of predictive success since it is a widely used concept in ecosystem modelling. The benefits and disadvantages with the r^2 value are probably better known than most alternative statistical measures (for example, predictive power, functional distance and/or confidence intervals).

There are uncertainties for all empirical mean values. One standard way of quantifying such uncertainties is, as mentioned, by means of the coefficient of variation, CV (CV = SD/MV; SD = standard deviation, MV = mean value). It is evident that the CV value of a given y-variable (like chlorophyll-a concentration in contexts of eutrophication or the Hg content in fish muscle in contexts of toxic contamination) is not a constant but a variable. The CV value for within ecosystem variability (CV_w) is always related to very complex climatological, biological, chemical and physical conditions. It has been shown (see Håkanson and Peters, 1995) that it is often possible to define a characteristic CV value for a given variable, like 0.35 for TP concentrations in lake water and 0.25 for Hg concentration in fish and chlorophyll.

The fit between modelled values and empirical data can be expressed by the r^2 value, and also by the difference between modelled values and empirical data by the mean difference (Diff = (M-E)/E; M = modelled value, E = empirical data). The r^2 value depends on the number of data and the range of the data in the regression (see appendix 8.1). When the number of data is small and the range narrow, one can preferably use the Diff value.

4.3.1. Variations within lakes and the highest r^2

If a lake model is tested against an independent set of measured data, the achieved r^2 value will depend on the uncertainty in the empirical y-value (the uncertainty in the y-direction) and on the structuring of the model, including which processes and model variables are accounted for, how this is done, and the empirical uncertainty of the model variables (the uncertainty in the x-direction; see fig. 4.4). Ecosystem models can never be expected to yield a high r^2 value if the target variable (y) cannot be empirically determined well, that is, if the coefficient of variation for within-system variations (CV_w) for the y-variable is high.

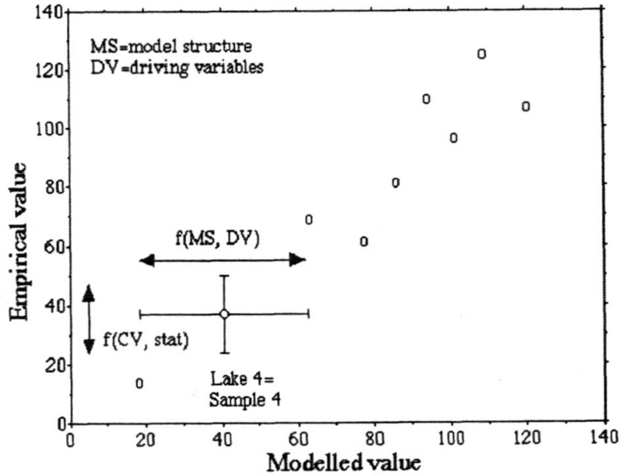

Fig. 4.4. Illustration of some fundamental concepts related to the question of "the highest possible r^2" of ecosystem models. The figure gives empirical data for the target variable on the y-axis and modelled values on the x-axis.

The within-lake variations causing uncertainties in the y-direction mainly depend on seasonal changes in climatological variables (such as temperature, light, winds, precipitation and bioproduction). This dependency is exemplified using monthly data on lake color from four years from a Swedish lake in fig. 4.5. For all lake variables there are also uncertainties related to sampling, transport, storage and analytical procedures. Generally, such methodological uncertainties are of the same order of magnitude as the natural variations (see Håkanson and Peters, 1995). Fig. 4.6 illustrates monthly within-lake variations (CV_w) for four lake variables, surface water temperatures (A), Secchi depths (B), TP concentrations (C) and color (D). The data in this figure, and many of the data used in the following regressions, are also described in more detail in appendix 8.3. All these lakes belong to one lake type, glacial lakes. There are about 83,000 lakes in Sweden and about 81,000 belong to this lake type; the same proportions ought to apply also for Finland, Russia, Canada and northern U.S.A. Most lakes on Earth belong to this lake type.

The data for color shown in fig. 4.5 for one lake are included in the data given in fig. 4.6C, where they are compared to similar data from many lakes. Fig. 4.6C shows that a characteristic CV_w for lake color varies around 0.2 (or 20%) and that there is a seasonal pattern. The CV_w is generally lower than 0.2 during the growing season and higher than 0.2 during fall and winter. Table 4.2 gives a compilation of characteristic CV_w values for many lake variables. These CV_w values regulate the highest r^2 value that one in practice can expect when modelled values are compared to empirical data (see eq. 4.1).

An approach to estimate the highest reference r^2, r_r^2, has been presented by Håkanson (1999).

$$r_r^2 = 1 - 0.66 \cdot CV_w^2 \qquad (4.1)$$

Where CV_w is the characteristic within-lake variability for the given y-variable. The equation is graphically shown in fig. 4.7. It is valid for actual (i.e., non-transformed) y-values. These r_r^2 values may be used as reference values for the following regressions. It can be noted from table 4.2 that r_r^2 for Secchi depth is 0.99, for TP it is 0.92, for color 0.97 and for pH 1.0. Note that pH is by definition a logarithmic value.

To conclude: the given CV_w values and the corresponding r_r^2 values will be used in the following to set the limits for the presented r^2 values in the regressions where the aim is to rank the role of lake morphometric parameters in predicting important lake variables.

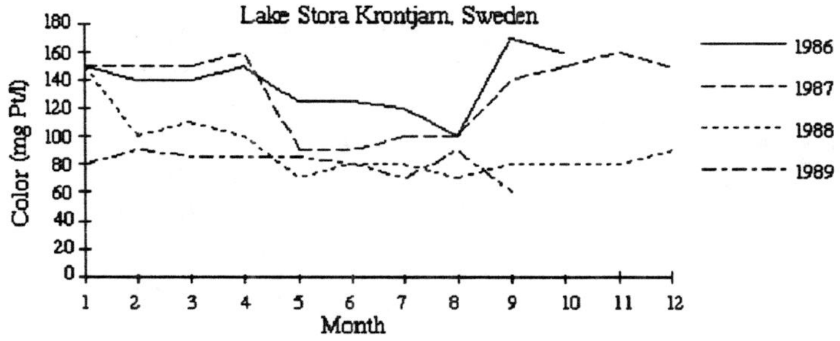

Fig. 4.5. Temporal variations in color in Lake Stora Krontjarn, Sweden, based on monthly data from 1986, 1987, 1988 and 1989 (data from Håkanson and Peters, 1995).

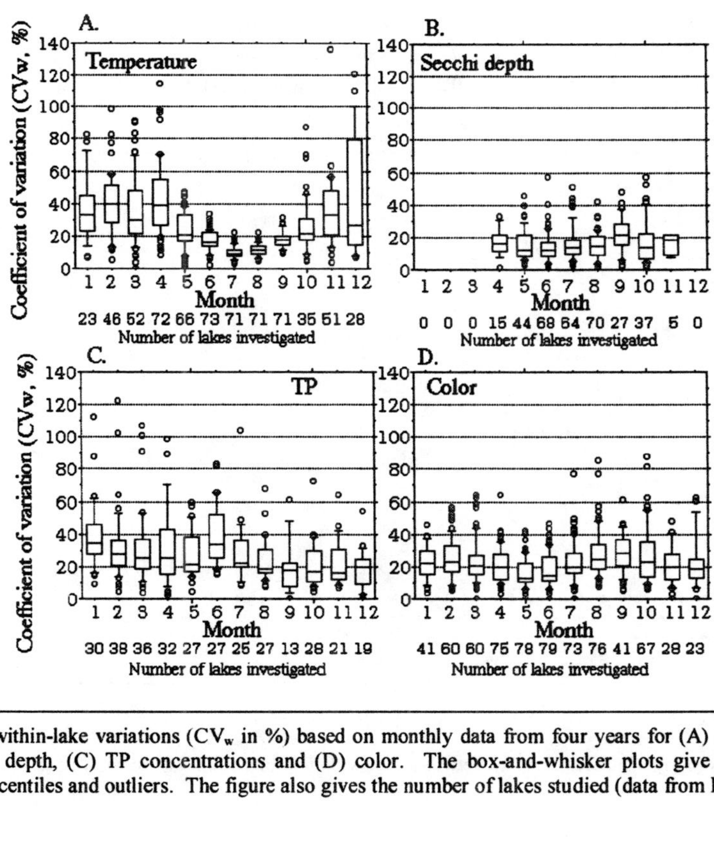

Fig. 4.6. Coefficient of within-lake variations (CV_w in %) based on monthly data from four years for (A) surface water temperatures, (B) Secchi depth, (C) TP concentrations and (D) color. The box-and-whisker plots give the medians, quartiles, 10th and 90th percentiles and outliers. The figure also gives the number of lakes studied (data from Håkanson and Peters, 1995).

Table 4.2. Compilation of characteristic CV_w values for different types of lake variables (from Håkanson, 2000) and the corresponding highest r^2 values (r_r^2).

	CV_w	r_r^2		CV	r_r^2
Lake variables			**Sedimentological variables**		
Lake area (A)	0.01	1.00	Percent ET areas (ET)	0.05	1.00
Mean depth (D_{mv})	0.01	1.00	Suspended particulate matter (SPM)	0.2	0.97
Maximum depth (D_{max})	0.01	1.00	Mean water content for E areas	0.3	0.94
Volume (V)	0.01	1.00	Mean water content for T areas	0.2	0.97
Theoretical water retention time (T)	0.1	0.99	Mean water content for A areas	0.05	1.00
			Mean bulk density for E areas	0.1	0.99
Water chemical variables			Mean bulk density for T areas	0.1	0.99
pH	0.05	1.00	Mean bulk density for A areas	0.02	1.00
Conductivity (cond)	0.1	0.99	Mean organic content for E areas	0.5	0.84
Ca concentration (Ca)	0.12	0.99	Mean organic content for T areas	0.5	0.84
Hardness (CaMg)	0.12	0.99	Mean organic content for A areas	0.1	0.99
K concentration (K)	0.2	0.97	Mean TP concentration for E areas	0.5	0.84
Color (Col)	0.2	0.97	Mean TP concentration for T areas	0.75	0.63
Fe concentration (Fe)	0.25	0.96	Mean TP concentration for A areas	0.35	0.92
Total-P concentration (TP)	0.35	0.92	Fall velocities	0.5	0.84
Alkalinity (alk)	0.35	0.92	Age of A sediments	0.5	0.84
			Age of ET sediments	0.5	0.84
Lake management variables			Diffusion rates	0.5	0.84
Secchi depth (Sec)	0.15	0.99			
Chlorophyll-a concentration (Chl)	0.25	0.96	**Climatological variables**		
Hg. concentration in lake fish	0.25	0.96	Annual runoff rates	0.1	0.99
^{137}Cs concentration in lake fish	0.22	0.97	Annual precipitation	0.1	0.99
			Temperatures	0.2	0.97

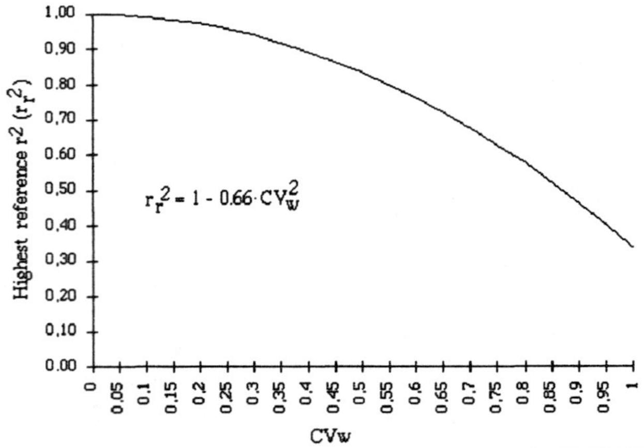

Fig. 4.7. Illustration of the relationship between the coefficient of within-lake variability (CV_w) and the highest reference r^2 (r_r^2).

4.3.2. Variations among and within lakes

The question concerning variability "within and among" is fundamental for issues related to representative values of lake variables: What does the lake sample represent? Which factors influence observed variations?

The problem addressed in this section is illustrated in table 4.3. The variability of any lake variable within an ecosystem, CV_w, is calculated from time-series data from the given lake: CV_w in table 4.3 is 0.72 for lake 1. The CV expressing variations among lakes, CV_a, is defined from the coefficient of variation of data from different lakes for a given period: It is 1.01, in the top row in the CV_a column in table 4.3.

Table 4.3. Definition of variability (= coefficient of variation) within, CV_w, and among, CV_a, lakes, for a hypothetical lake variable.

Period	Lake 1	Lake 2	Lake 3	Lake 4	Lake 5	MV_a	SD_a	CV_a
1	4.2	8.5	1.5	15.9	0.7	6.16	6.24	1.01
2	4.0	4.2	0.8	18.8	9.6	7.48	7.07	0.95
3	6.8	3.1	1.5	16.5	11.8	7.94	6.21	0.78
4	15.4	1.5	10.0	15.7	14.8	11.48	6.04	0.53
5	20.1	1.4	11.6	11.0	18.0	12.42	7.32	0.79
MV_w	10.1	3.7	5.1	15.6	11.0			
SD_w	7.3	2.9	5.3	2.8	6.6			
CV_w	0.72	0.78	1.04	0.18	0.60			

Fig. 4.8 gives a compilation of such CV_w and CV_a values for seven standard lake variables (surface water temperature, pH, alkalinity, TP, color, Secchi depth and conductivity) based on monthly data from four years in 24 Swedish lakes. As expected from this geographically restricted area of the world, there are no significant differences between CV_w and CV_a for surface water temperature. The median CV value for temperature is even higher within lakes than among them. In contrast, the median CV values for pH, color, TP, Secchi depth and conductivity are much higher among than within lakes. In this example, the greatest difference in CV_a and CV_w exists for conductivity; the smallest (except temperature) for pH and alkalinity.

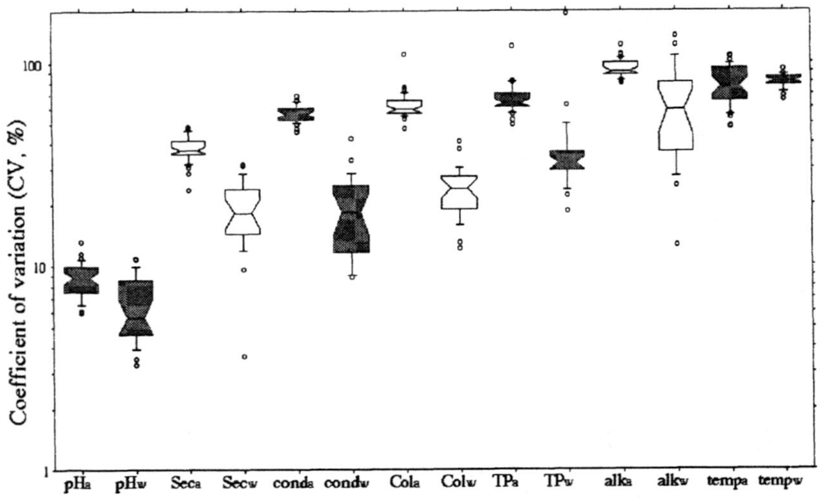

Fig. 4.8. Box-and-whisker plots (showing 10th, 25th, 50th, 75th and 90th percentiles and outliers) of coefficients of variation (CV) for variability within (w) and among (a) lakes for lake water temperature (°C), pH, TP (µg/l), color (Col, mg Pt/l), Secchi depth (m) and conductivity (m S/m). Based on monthly data from 24 lakes (data from Håkanson and Peters, 1995).

If the difference in CV_a and CV_w is large, or the ratio high (see fig. 4.9A and table 4.4), very good regression models for water variables (such as conductivity and hardness) can be derived based on parameters describing the catchment area. Fig. 4.9B gives the CV_a/CV_w ratio on the y-axis and the maximum obtained r^2 value in regressions based only on catchment area parameters on the x-axis. Fig. 4.9C gives a regression between the CV_a/CV_w ratio and the maximum r^2 values from regressions based only on lake morphometric parameters. One can note that the r^2 value is low and that lake morphometric parameters are not linked to the CV_a/CV_w ratio in the same way as catchment area parameters. The reason for this will be discussed in the next section. The regressions used in fig. 4.9 will be presented in a following part of this chapter.

From this section, one can conclude that for some lake variables (conductivity, iron, color, hardness and calcium), the conditions in the catchment area determine to a very high degree the characteristic long-term lake value; and for those variables, lake morphometry plays a relatively small role (see fig. 4.9C).

Table 4.4. Results for various lake variables giving the r^2 values obtained when these variables have been regressed against lake morphometric parameters, catchment area parameters, lake morphometric plus catchment area parameters; and the ratio between the CV values for these variables among and within lakes.

Variable	r^2 vs lake morphometry	r^2 vs catchment	r^2 vs lake + catchment	CV_{among}/CV_{within}
Temperature	•	•	•	0.98
pH_{12}	0.07	0.46	0.46	1.55
alk_{12}	0	0.34	0.34	1.69
TP_{12}	0.15	0.32	0.37	1.91
Ca_{36}	0.08	0.55	0.62	2.07
Sec_{36}	0.51	0.46	0.65	2.12
Col_{36}	0.28	0.69	0.68	2.45
$CaMg_{12}$	0.13	0.67	0.71	2.51
Fe_{36}	0.19	0.70	0.71	2.65
$cond_{12}$	.15	0.69	0.72	3.26

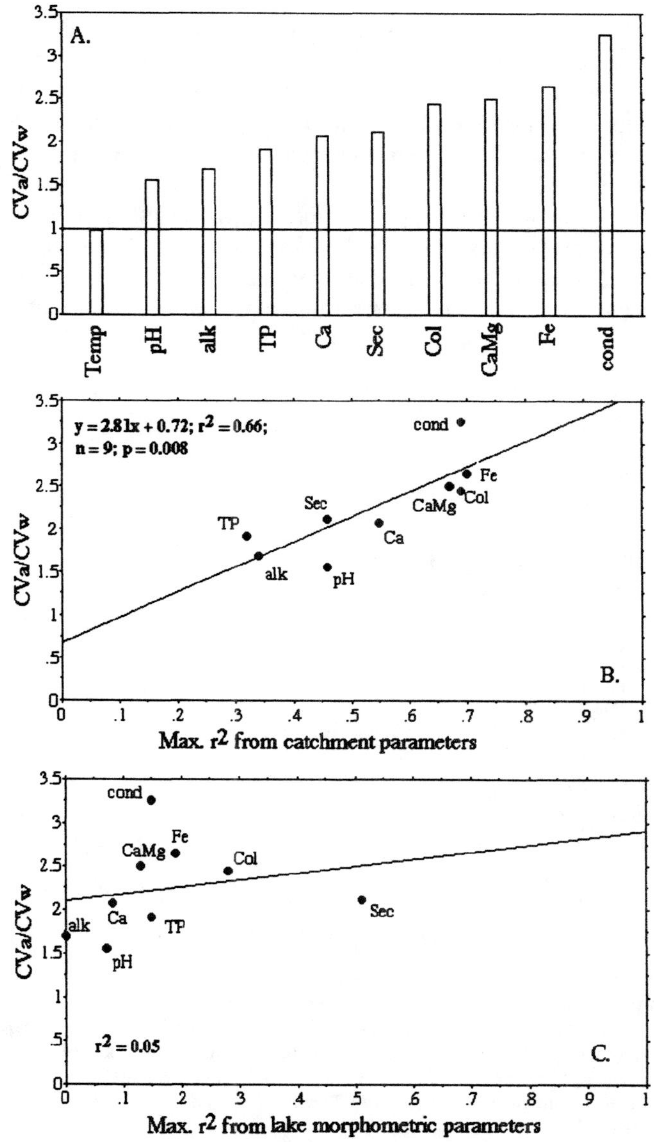

Fig. 4.9. A. The ratio of among lake variability to within lake variability for ten standard variables based on monthly data from 24 lakes.; B. The relationship between the "Among/Within" ratio and the maximum r^2 values obtained for empirical regression models based on catchment area parameters for these water chemical variables (alk_{12} means mean annual alkalinity, i.e., the value is based on data from 12 months, etc.); C. The same the "Among/Within" ratio versus the maximum r^2 values obtained for empirical regression models based on lake morphometric parameters.

4.3.3. Conservative and reactive substances and basic mass-balances

On the other hand, lake morphometry plays a key role for variables that settle out or in other ways participate in biological or chemical lake processes. This role is indicated in fig. 4.10, which gives the maximum r^2 values in regressions based only on lake morphometric parameters on the y-axis and the sedimentation rates for several variables in Lake Ekoln, Sweden on the x-axis. The r^2 values in fig. 4.10 will be presented later in this chapter and the logic and relevance of the results will be explained in this section.

The sedimentation rates in fig. 4.10 emanate from empirical measurements from a 10-year period of both input and output to this lake. Table 4.5 gives a compilation of sedimentation rates and fall velocities for many variables included in that study (see Håkanson and Jansson, 1983). To understand the results shown fig. 4.10, one must also discuss basic components of mass-balance calculations.

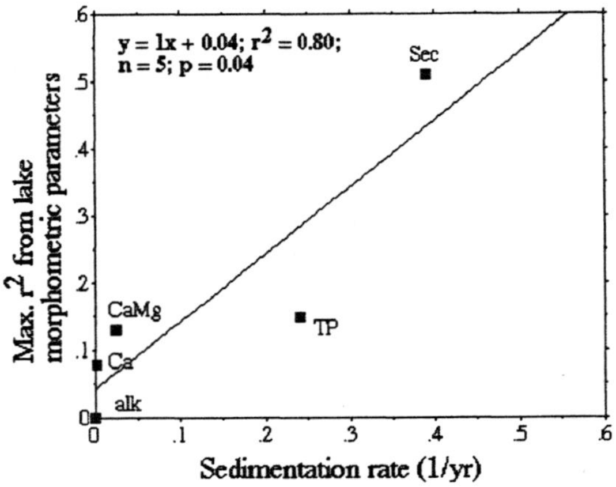

Fig. 4.10. The relationship between the maximum r^2 values obtained for empirical regression models based on lake morphometric parameters versus the sedimentation rate (1/yr) for five lake variables (Secchi depth, total phosphorus concentrations, hardness = CaMg, calcium concentrations and alkalinity) using data for Lake Ekoln, Sweden. For this lake there has been a full mass-balance so that both the inflow and the outflow have been measured for the period 1967-1977. The theoretical lake water retention time for Lake Ekoln is 2.07 years. The lake area is 18.6 km^2 and the mean depth 19 m (data from Håkanson and Jansson, 1983).

Table 4.5. Sedimentation rates (1/yr) and fall velocities (m/yr) for standard lake variables based on measured data on inflow and outflow for Lake Ekoln, Sweden (area = 18.6 km^2; mean depth = 19 m), for the period 1967-1977 (from Håkanson and Jansson, 1983).

Variable		Sedimentation Rate	Fall Velocity
Conservative	Alkalinity	0	0
	Chloride	0	0
	Calcium	0	0
	Sodium	0.005	0.1
	Magnesium	0.048	0.9
	Sulfate	0.048	0.9
	Potassium	0.048	0.9
	Color	0.082	1.6
	Organic nitrogen	0.15	2.9
	Phosphate-P	0.16	3.0
	Total-N	0.22	4.2
	Nitrate-N	0.22	4.2
	Total-P	0.24	4.6
	Nitrite-N	0.25	4.8
	Silicon	0.26	4.9
	Particulate-P	0.32	6.1
	Suspended part. matter	0.39	7.4
	Secchi depth	.40	7.6
Reactive	Ammonia-N	0.41	7.8

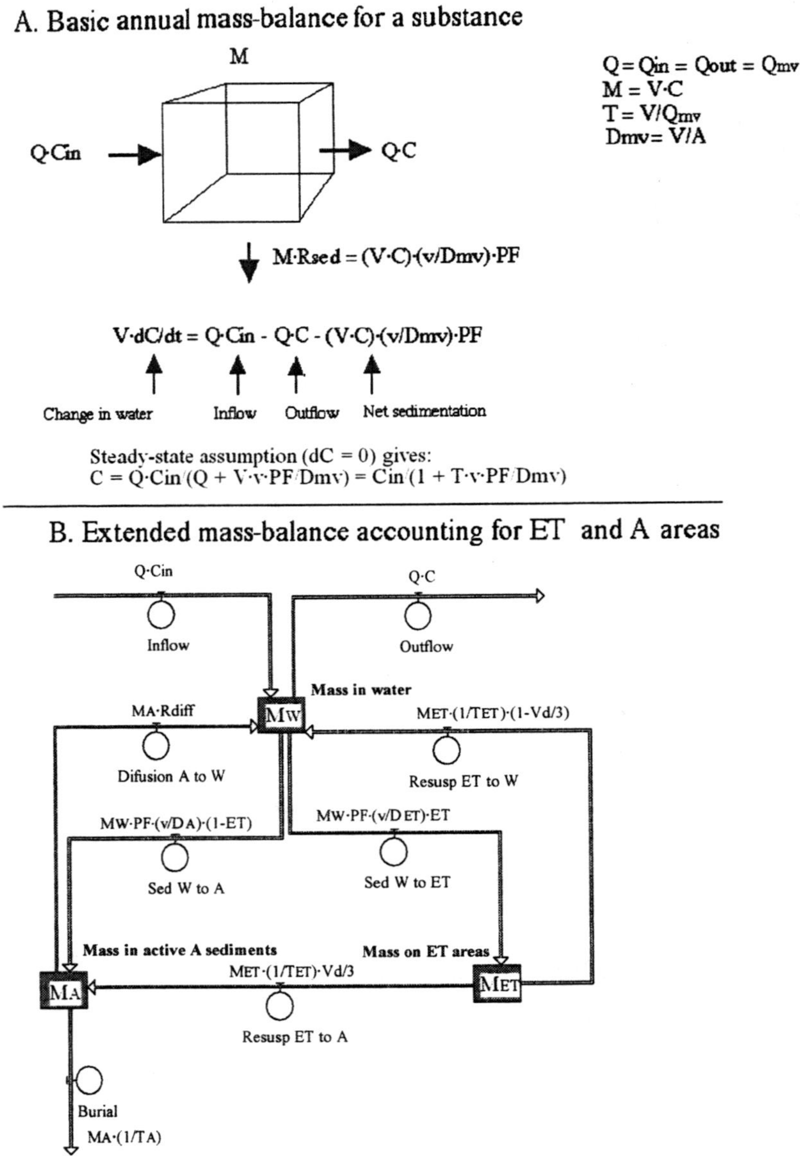

Fig. 4.11. A. Illustration of the basic annual mass-balance for a substance showing input, output and net sedimentation; as well as commonly used abbreviations; and the steady-state solution expressing C as a function of C_{in}, T, v, PF and D_{mv}; where C is the mean lake concentration, C_{in} the mean tributary concentration, T the theoretical lake water retention time, v the settling velocity, PF the particulate fraction and D_{mv} the mean depth. B. Shows an extended mass-balance approach also accounting for internal loading. This approach quantifies sedimentation on ET areas and on A areas, resuspension from ET areas back to lake water and to A areas, diffusion from A areas to lake water and burial.

The basic mass-balance model for the annual flow of matter, or in this example total phosphorus (TP), to and from a lake may be described by the differential equation illustrated in fig. 4.11A. The lake is envisioned as a reactor tank with complete mixing during the calculation time (dt). The simplest way of solving this equation is to assume steady-state conditions. That means that one conserves mass and puts $Q_{in} = Q_{out} = Q$.

$$C = Q_{mv} \cdot C_{in} / (Q_{mv} + V \cdot v \cdot PF / D_{mv}) \quad (4.2)$$

Where

C = the mean concentration of the substance in the lake (mg/m^3);
C_{in} = the mean concentration of the substance in the tributary (mg/m^3);
Q_{mv} = the mean annual tributary water discharge to the lake (m^3/yr);
V = the lake volume (generally m^3 or km^3);
v = the fall velocity of the particulate phase (m/yr or similar units);
D_{mv} = the mean depth of the lake (m);
PF = the particulate fraction (dimensionless); the only fraction that can settle out by gravity; PF = 1-

DF, where DF is the dissolved fraction, which is generally approximated to be the bioavailable fraction, i.e., the fraction available for direct uptake by biota in the lake water; $PF = C_P/C_T$, where C_P is the particulate concentration (mg/m^3 or similar units) and C_T the total concentration (mg/m^3 or similar units).

The theoretical lake water retention time (or residence or turnover time, T, in time units; generally years) is defined as the ratio between the lake volume and the mean annual water discharge, or $T = V/Q_{mv}$. T is the time it takes to fill a compartment of volume V if the water discharge to the lake is given by Q_{mv}. Eq. 4.2 can be rewritten as:

$$C = C_{in} / (1 + T \cdot v \cdot PF / D_{mv}) \quad (4.3)$$

The sedimentation rate (R_{sed}) is:

$$R_{sed} = (v / D_{mv}) \cdot PF \quad (4.4)$$

The sedimentation rate of the given substance C in the lake has, like all rates, the dimension 1/time, that is, per time unit (here 1/year). R_{sed} expresses how much of the mass (in g) of the substance settles during the given time; if R_{sed} is 0.1, it means that 10% of the mass settles per year, etc.

Eq. 4.3 is graphically illustrated in fig. 4.12, which highlights the importance of the particulate fraction in predicting lake TP concentrations (fig. 4.12A), and the mean depth and the theoretical lake water retention time in predicting lake TP concentrations (fig. 4.12B).

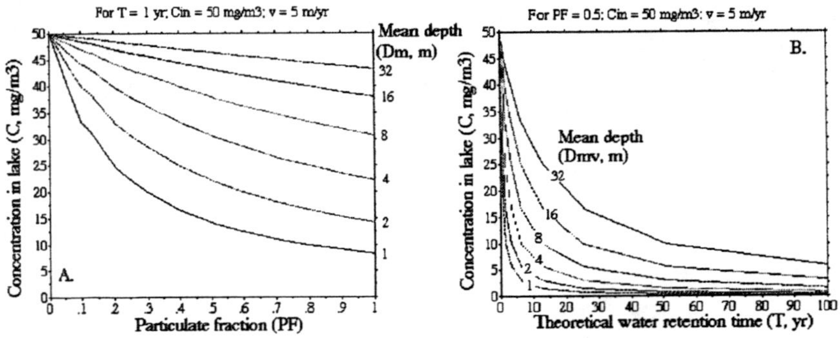

Fig. 4.12. Illustration of the basic mass-balance model.
A:Gives the mean concentration in lake water as a function of the particulate fraction and the mean depth for a lake with a theoretical lake water retention time, T, of 1 yr, a mean annual tributary concentration, C_{in}, of 50 mg/m^3 and a fall velocity of the particulate matter of 5 m/yr.;
B: Gives the mean concentration in lake water as a function of the theoretical lake water retention time and the mean depth for a lake with a particulate fraction, PF, of 0.5, a mean annual tributary concentration, C_{in}, of 50 mg/m^3 and a fall velocity of the particulate matter of 5 m/yr.

Substances which only appear in dissolved phase, such as alkalinity, chloride and calcium in Lake Ekoln (see table 4.5) are generally referred to as conservative or non-reactive (substances which do not change, settle, evaporate or are being transformed in the lake); most allochthonous particles and pollutants, which are transported to the lake from the catchment, have a particulate fraction (PF) higher than zero and are distributed in a typical patterns with lobes of decreasing concentrations with distance from the source of pollution, or from the river mouth.

In the first load model for lakes, presented by Vollenweider (1968), the basic mass-balance equation was somewhat altered. Empirical data indicated that a better prediction of C for TP (C_{TP}) could be obtained by the following formula:

$$C_{TP} = C_{TPin}/(1+\sqrt{T}) \qquad (4.5)$$

In this expression, there is no sedimentation rate, R_{sed}, and instead of using $R_{sed}\cdot T$, one has $\sqrt{T} = T^{0.5}$, which gave better predictability when empirically tested. An even better prediction of C_{TP} can often be obtained by the following model (OECD, 1982; see fig. 4.13):

$$C_{TP} = 1.55\cdot(C_{TPin}/(1+\sqrt{T}))^{0.82} \qquad (4.6)$$

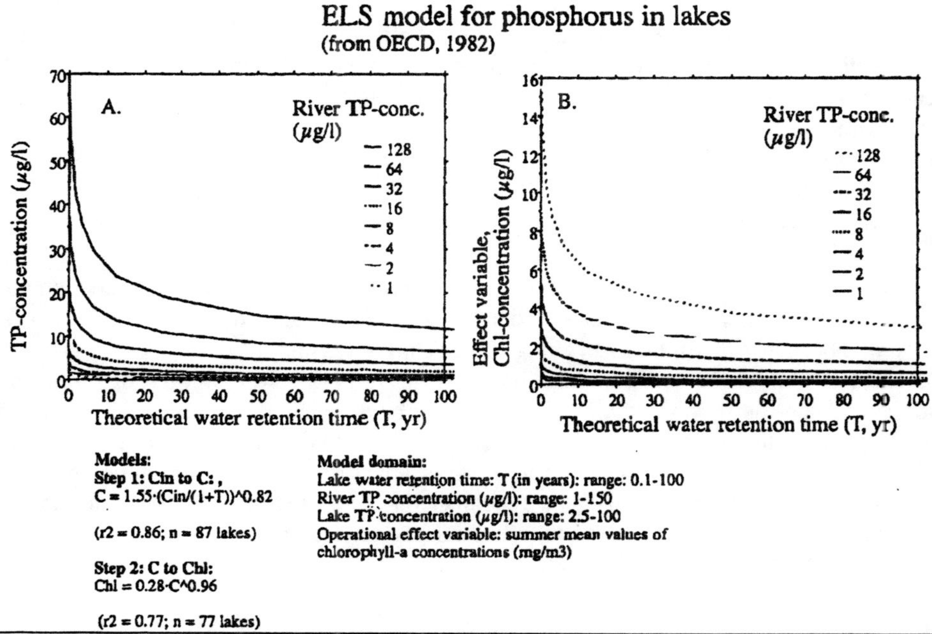

Fig. 4.13. Illustration of an effect-load-sensitivity model for phosphorus in lakes (the OECD model). The figure also gives the models and information on the model domain.
A: Mean annual TP concentrations as a function of the theoretical lake water retention time and mean annual tributary TP concentration.
B: Gives the operational effect variable used in many lake studies, the concentration of chlorophyll-a, as a function of the theoretical lake water retention time and mean annual tributary TP concentration.

It is based on the same driving parameters as the original Vollenweider model (C_{in} and T) but uses three empirical constants, 1.55 and the two exponents 0.5 (= √) and 0.82; C_{TP} and C_{TPin} represent mean annual concentrations.

If C_{TP} from eq. 4.6 is put into the first equation in table 4.1, one obtains the classical ELS model expressing mean summer chlorophyll concentrations as a function of mean tributary TP concentrations and theoretical lake water retention time (which is basically a morphometric parameter). This is shown in fig. 4.13.

Note that models of the OECD type:

- use mean annual values, hence they do not account for seasonal variations;
- do not account for internal loading, which is important in shallow, eutrophic lakes, the lake type for which the load models are needed the most;
- do not differentiate between bioavailable (≈ dissolved) and non-bioavailable (≈ particulate) fractions of phosphorus. Operationally, this separation is generally done by filtering through a 0.45 μm filter.

Meeuwig and Peters (1996) listed 17 models of the Vollenweider type for predicting mean annual lake TP from different empirical variables and morphometrical parameters. The reason for this multitude of models is that these models are, evidently, often too crude and uncertain to be useful for many lake management applications in individual lakes. Many models of the OECD type have been developed for certain regions or for specific lake types and often do not work well outside a relatively narrow range of lakes.

Mass-balance modelling can be done in different ways (different scales, target variables, driving variables and structures). Another more realistic modelling approach, which accounts for internal loading and relates more to lake morphometry, is illustrated in fig. 4.11B. This modelling approach will be discussed in more detail in chapter 5. It gives fluxes to and from the following compartments:

- Lake water.
- ET areas (defined by the area above the wave base, see fig. 3.23)
- Accumulation areas (the areas below the wave base).

Fig. 4.11B also gives the fluxes for the following processes: Inflow, outflow, sedimentation on ET areas and on A areas, resuspension from ET areas to water and from ET areas to A areas, diffusion and burial. All will be treated in greater detail in chapter 5.

Substances with high sedimentation rates (R_{sed} values) settle rapidly near the point of discharge, and substances with small R_{sed} values may be distributed over much larger areas. Rsed is related to the settling velocity, v ($v = D \cdot R_{sed}$, where D is the depth or distance through which the particle sinks during the given time interval). Given R_{sed} or v, one can model or predict where high and low concentrations are likely to appear in water and sediments. This information is the key to predicting where high and low ecological effects may appear.

Stokes' law expresses the settling velocity (v) as:

$$v = (d_w - d_p \cdot g \cdot L_{pd}^2)/(18 \cdot AV \cdot FR) \qquad (4.7)$$

Where

- v = the settling velocity (usually in cm/s or m/month);
- d_p = the particle density (usually in g dw/cm^3; g dw = gram dry weight);
- d_w = the density of the lake water (often set to 1 g ww/cm^3; g ww = gram wet weight);
- g = the acceleration due to gravity (980.6 cm/sec^2);
- L_{pd} = the particle diameter (in m, cm or mm);
- AV = the coefficient of absolute viscosity (obtained from standard tables; 0.01 poise at 20°C);
- FR = the coefficient of form resistance (set to 1 for spheres; Hutchinson, 1967).

Stokes' law (Stokes, 1851) is depicted in fig. 4.14. The behavior of material that follows Stokes' law (particles with a diameter between about 0.01 and 0.0001 cm) differs from that of the coarser friction material and from that of the still finer material. The sedimentological behavior of the friction material is closely linked to the grain size of the individual particles (Einstein, 1950; Allen, 1970); the sedimentological behavior of the very fine collodial materials is governed by Brownian motion (see fig. 4.14). These latter particles are so small that they will not settle individually, but will do so if they form larger flocs or aggregates that are dense enough to settle according to Stokes' law (Kranck, 1973, 1979; Lick et al., 1992). The cohesive materials that follow Stokes' law are very important, since they have a great affinity for pollutants. This group includes many types of detritus, humic substances and plankton. All play significant roles in aquatic ecosystems (Salomons and Förstner, 1984). So, related to lake functional aspects, there are basically three major categories of matter in lake water: (1) the particulate fraction (PF), which is the only fraction settling by gravitation, (2) the dissolved fraction, which is the most important one for direct bio-uptake by, for example, phytoplankton and often referred to as the bioavailable fraction, and (3) the colloidal fraction, which is neither subject to direct uptake nor to sedimentation because the particles are too small. The particulate fraction and the colloidal fraction may be subject to mineralization by bacterioplankton.

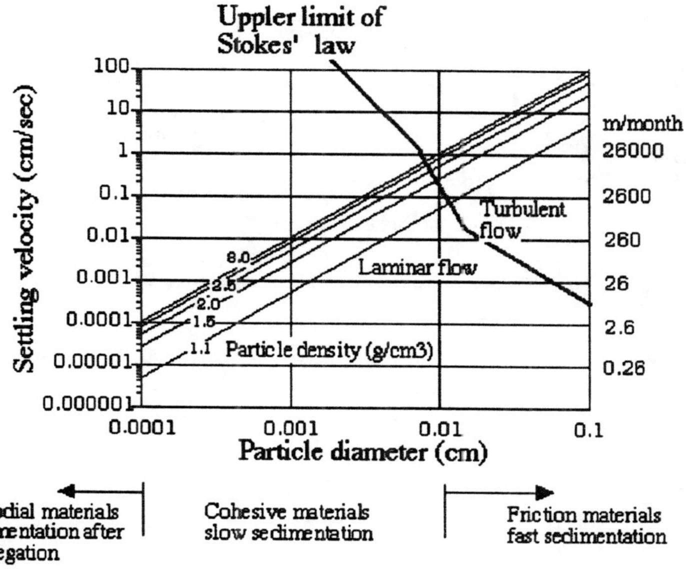

Fig. 4.14. The relationship between the settling velocity (v; of spherical particles) in water, particle diameter and particle density (at 20°C) as given by Stokes' law.

The settling velocity (v) of a given particle, aggregate, or particulate pollutant, and its distribution in a lake, depend on the density, size and form of the particle (and on the hydrodynamics of the flow of water in the lake; see fig. 4.15). If the particle density, d_p, is close to 1, if the coefficient of form resistance is large, and if the diameter is small, the settling velocity and the sedimentation rate may be very small. If the sedimentation rate is close to 0, the particle or aggregate is conservative in the sense that it may not be deposited in the lake. In many lakes (see table 4.5), Cl, Ca and alkalinity are typically conservative substances, and color, organic matter, total P, Si, particulate P and suspended matter are more reactive.

Fig. 4.16 illustrates some important principles regarding PF values for different metals and radionuclides. The information of cation exchangeability illustrated for Hg, Pb, Cd, Cu, Zn, Cr and Ni in fig. 4.16 comes from the literature (Förstner and Wittmann, 1981; Salomons and Förstner, 1984; Lindström, 2000); and the information on PF for radiocesium and radiostrontium and how the PF value for radiocesium depends on the potassium concentration in the lake water and the PF value for radiocesium on the calcium concentration comes from (IAEA, 2000 and Håkanson, 2000). Elements with high PF values are likely to settle out in lakes and be more influenced by characteristics of lake size and form regulating internal fluxes than elements which appear with high dissolved fractions.

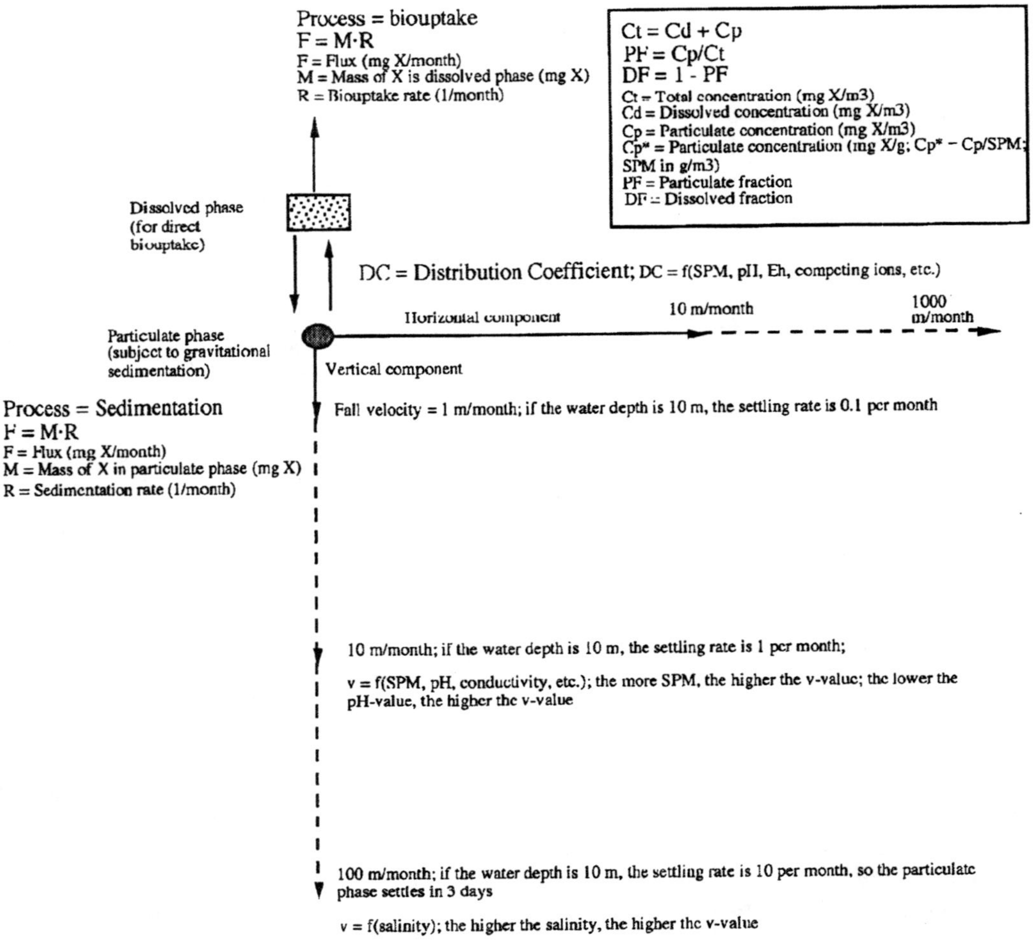

Fig. 4.15. Illustration of the basic distribution coefficient regulating how much a given substance or pollutant that is present in dissolved or particulate forms. Only the particulate fraction can settle by gravity. Note that particles rarely settle in a simple vertical manner in natural aquatic systems. The horizontal component is generally much greater than the vertical component. The figure also gives basic definitions of fluxes and distribution coefficients.

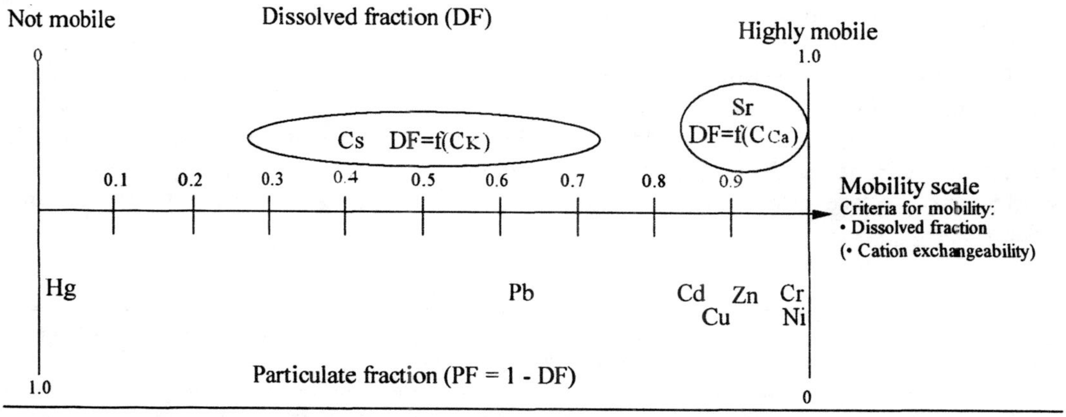

Fig. 4.16. Illustration of particulate fraction (PF) for metals and radionuclides.

From this, one can conclude that lakes of different size and form will not have the same sedimentation rates for different substances since the sedimentation rate is basically defined by the ratio between the fall velocity (v in m/yr) divided by the mean lake depth (D_{mv} in m). It is interesting to note the significant, logical and positive relationship in fig. 4.10 between the sedimentation rate, which expresses how the variable is retained in the system [the higher the rate the more of the substance is retained in the lake (if the rate is 0.4, it means that 40% of the yearly input stays in the lake)] and the r^2 from the regression models based on lake morphometric parameters. Conservative substances, like alkalinity, which do not settle out by gravity in this lake or most lakes, are not really influenced by lake morphometry. These substances appear in dissolved and not in particulate form. Reactive substances with sedimentation rates higher than zero, on the other hand, may be strongly influenced by lake morphometry.

The main conclusion is that lake morphometry is very important if one would like to understand how reactive substances are distributed in lakes. The same applies for all substances that appear with PF values higher than zero. As stressed, the settling rate also depends on the grain sizes, the flocculation and the density of the particles/flocs.

4.3.4. Internal correlations among lake morphometric parameters

Table 4.6 gives a correlation matrix showing internal co-variations amont 15 morphometric parameters and fig. 4.17 shows nine selected regressions between morphometric parameters. The idea here is to demonstrate internal correlations correlations among these morphometric parameters. Parameters showing high internal correlations, such as those belonging to the same cluster, may replace one another in regression models, so it is interesting to clarify such relationships.

Table 4.6. Correlation matrix giving correlation coefficients, r, amount lake morphometric parameters based on data from 95 glacial boreal lakes (data from Håkanson and Peters, 1995).

	Alt	V_d	log(A)	log(ADA)	log(ADA/A)	log(D_{mv})	log(D_{max})	log(DR)	log(D_{rel})	log(ET)	log(L_d)	log(RDA)	log(Q_{mv})	log(T)
Alt	1													
V_d	-0.26	1												
log(A)	0.00	-0.33	1											
log(ADA)	-0.14	-0.14	0.65	1										
log(ADA/A)	-0.17	0.22	-0.40	0.44	1									
log(D_{mv})	0.14	-0.47	0.30	0.16	-0.16	1								
log(D_{max})	0.21	-0.73	0.37	0.19	-0.21	0.94	1							
log(DR)	-0.11	0.13	0.58	0.39	-0.22	-0.59	-0.49	1						
log(D_{rel})	0.21	-0.48	-0.39	-0.30	0.09	0.71	0.71	-0.92	1					
log(ET)	-0.25	0.47	-0.21	0.06	0.33	-0.47	-0.54	0.21	-0.37	1				
log(L_d)	-0.03	-0.35	0.28	0.18	-0.12	0.06	0.19	0.20	-0.03	-0.18	1			
log(RDA)	0.20	-0.03	-0.43	-0.29	0.16	0.18	0.15	-0.52	0.47	-0.18	-0.39	1		
log(Q_{mv})	-0.10	-0.18	0.67	0.98	0.40	0.22	0.25	0.36	-0.25	-0.01	0.19	-0.27	1	
log(T)	0.14	-0.40	0.51	-0.21	-0.86	0.59	0.61	-0.06	0.21	-0.43	0.12	-0.08	-0.17	1
log(V)	0.05	-0.46	0.92	0.59	-0.38	0.65	0.68	0.22	-0.02	-0.36	0.25	-0.27	0.63	0.65

From fig. 4.17, one can note that:

- There is a strong ($r^2 = 0.88$), logical correlation between the mean depth (D_{mv}) and the maximum depth (D_{max}; fig. 4.17A).
- There is also a strong ($r^2 = 0.97$) and very important relationship between the catchment area (ADA in km^2) and the mean annual water discharge (Q_{mv} in m^3/s; fig. 4.17B). The regression line in fig. 4.17B may also be expressed as $Q_{mv} = 0.0091 \cdot ADA^{0.93}$. Since Q_{mv} is physically defined by $Q_{mv} = SR \cdot ADA$, where SR is the specific runoff of water from land in $m^3/km^2 \cdot s$, one can conclude that SR on average is about 0.009 for these catchments.
- There is also a logical and strong ($r^2 = 0.85$) correlation between lake area (A) and volume (V; fig. 4.17C).
- Since the mean water discharge (Q_{mv}) may be predicted very well from ADA and since there is a strong relationship between volume and area, one can expect that the theoretical water retention time (T), which is defined as $T = V/Q_{mv}$, should be highly correlated to the catchment area to lake area ratio (ADA/A). This high correlation is also the case, as shown in fig. 4.17D ($r^2 = 0.73$).

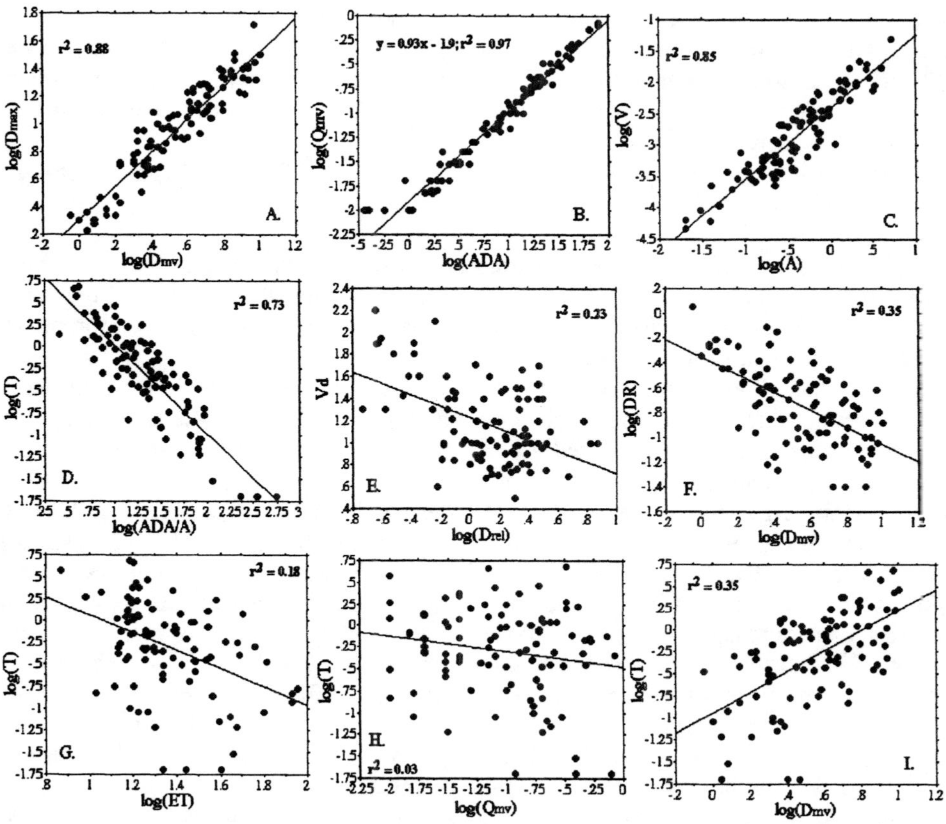

Fig. 4.17. Nine scatter plots illustrating the statistical relationships (coefficients of determination, r^2) between important morphometrical parameters in contexts of lake ecosystem functioning based on data from 95 boreal glacial lakes (data from Håkanson and Peters, 1995).
A. Mean depth (D_{mv} in m) versus maximum depth (D_{max} in m);
B. Area of drainage area (ADA in km^2) versus mean annual water discharge (Q_{mv} in m^3/s);
C. lake area (A in km^2) versus lake volume (V in km^3);
D. the catchment to lake ratio (ADA/A) versus theoretical lake water retention time (T in years);
E. relative depth (D_{rel}) versus volume development (V_d);
F. mean depth (D_{mv}) versus dynamic ratio (DR);
G. bottom areas of erosion and transport (ET in % of lake area) versus theoretical lake water retention time (T in years);
H. mean annual water discharge (Q_{mv} in m^3/s) versus theoretical lake water retention time (T); and mean depth (D_{mv}) versus theoretical lake water retention time (T).

- One might expect that there should also be a strong relationship between the volume development (the form factor, V_d) and the relative depth (D_{rel}) since both these morphometric parameters express aspects of lake form, but this is not the case (fig. 4.17E; $r^2 = 0.23$). The reason is that they express different elements of lake form ($V_d = 3 \cdot D_{mv}/D_{max}$ and $D_{rel} = D_{max} \cdot \sqrt{\pi}/(20.0 \cdot \sqrt{A})$).
- From the definition of the dynamic ratio ($DR = \sqrt{A}/D_{mv}$), one might also expect that DR should be highly related to the mean depth (D_{mv}), but that is not the case (fig. 4.17F; $r^2 = 0.35$). These two parameters also express different form elements.
- The theoretical lake water retention time (T) and the ET areas (ET) are both fundamental components of dynamic mass-balance models since T regulates the retention time of water, and everything dissolved and suspended in the water, and ET (see fig. 4.11B) regulates resuspension and internal loading. Fig. 4.17G shows that there is only a weak

correlation between these two variables ($r^2 = 0.18$) for these lakes.

- Fig. 4.17H shows that there does not exist any significant correlation ($r^2 = 0.03$) between the theoretical lake water retention time (T) and water discharge (Q_{mv}) in spite of the fact that T is defined as V/Q_{mv}. Two lakes can have the same inflow of water but very different volumes!
- There is a stronger relationship between T ($= V/Q_{mv}$) and D_{mv} ($=V/A$) than between T and Q_{mv} (fig. 4.17I; $r^2 = 0.35$).

Table 4.7 gives a compilation of cluster variables ranked according to correlation coefficients in three categories, (1) variables generally showing r values between 0.5 and 0.7, (2) between 0.7 and 0.9 and (3) r values higher than 0.9. These correlations are shown for lake morphometric variables (A), catchment area parameters (B) and water variables (C).

One can conclude from these correlations that different lake morphometric parameters are defined to express different elements of size and form, and that lakes can take many shapes. All lakes are individuals and react individually to changes in loading of nutrients or toxins. A fundamental part of that individuality is lake morphometry.

Fig. 4.18 summarizes some other important concepts in statistical analyses of lake data. Assume that one has an important y-variable that one would like to describe, understand and predict. Hundreds of interrelated x-variables could potentially influence the y-value, and (in lake studies) many of these also vary in time and space. Given a comprehensive data set with reliable empirical data, one could use a variety of statistical analyses (like correlations, regressions, factor analysis, stepwise multiple regression) of the untransformed and transformed x-variables to make a first ranking of the factors that appear to be the most powerful predictors. Assume that we obtain a model with three x-variables (x_1, x_2 and x_3), three slopes (a_1, a_2 and a_3) and an intercept (b). If one of the x-variables was lake pH, one could ask a number of general questions about pH. The same questions could be addressed to any water chemical x-variable.

- Since lake pH can vary seasonally in a lake, one must ask which pH value is most relevant. If one is to use the pH value for one month (pH_1), then which month should be used? Is a mean value for three months better, and if so, which months? Would a mean taken over 6, 12 or 24 months be better still? In short, which time period, which variant, should be used in relation to the y-variable one wants to predict?
- Which transformation should be used? Is y normally distributed or not? Is x (here pH) normally distributed? And if not, which transformation would produce a normal frequency distribution? $\sqrt{x}$, log(x), log(x+1), 1/x, or what? What is the likely frequency distribution for the entire lake population (of this lake type)? What is the frequency distribution for the lakes in the given sample of lakes?
- If x (pH) gives the highest r^2 value according to the chosen statistical method, does that mean that x is causally related to y? Is another member of the same cluster more likely to be the real causal variable? Is this fact hidden by poor empirical data for the other variable? Or by something else? For example, because lake pH belongs to a cluster of related variables, like alkalinity and hardness, one should ask if these variables should be used in the predictive model instead of pH.

These problems are not trivial ones. The correct answer to these questions could make a great difference to the predictive power of the model and to our understanding of causes for variations in lake variables.

Table 4.7. Compilation of clusters for lake morphometric parameters (A), catchment parameters (B) and water variables (C) (from Håkanson and Peters, 1995).

	Cluster parameters		
	$r>0.90$	$0.70<r<0.90$	$0.50<r<0.70$
A. Morphometric parameters			
Area (A)	V		DR, T, (ADA, Q)
Volume (V)	A		D_{max}, D_{mv}, T, (ADA, Q)
D_{max}	D_{mv}	V_d, D_{rel}	V, T, ET, DR
D_{mv}	D_{max}		D_{rel}, V, T, DR
V_d		D_{max}	
DR	D_{rel}		D_{mv}, A RDA
ET			D_{max}
D_{rel}	DR	D_{max}, D_{mv}	
L_d			
T		A/ADA	V, D_{max}, D_{mv}, A
B. Catchment parameters			
Drainage area (ADA)	Q		(A, V)
Relief (RDA)			DR
Lake%			
Mire%		Forest%	
Forest%		Mire%	
Open land%			
Rock%		Till%	
Till%		Rock%	
Q	ADA		A, V
area/ADA		T	
C. Water variables (annual mean values)			
pH_{12}		alk_{12}	Ca_{12}, $CaMg_{12}$
alk_{12}		pH_{12}	Ca_{12}, $CaMg_{12}$
$cond_{12}$		$CaMg_{12}$, Ca_{12}	
$Color_{12}$		$Secchi_{12}$, Fe_{12}	TP_{12}, ($Temp_{12}$)
$Secchi_{12}$		$Color_{12}$	TP_{12}, Fe_{12}
Fe_{12}		$Color_{12}$	TP_{12}, Sec12
TP_{12}			$Secchi_{12}$, $Color_{12}$, Fe_{12}
Ca_{12}	$CaMg_{12}$	$cond_{12}$	alk_{12}, pH_{12}
$CaMg_{12}$	Ca_{12}		$cond_{12}$, alk_{12}, pH_{12}
$Temp_{12}$			$Color_{12}$

Different **variants** of x1, x2 and x3, e.g.:
• yearly lake mean value
• mean lake value from winter (6 months)
• mean lake value from the growing season (3 months)
• mean lake value for a single month
• mean lake value from several samplings during one day
• site mean value or area (= lake) mean value

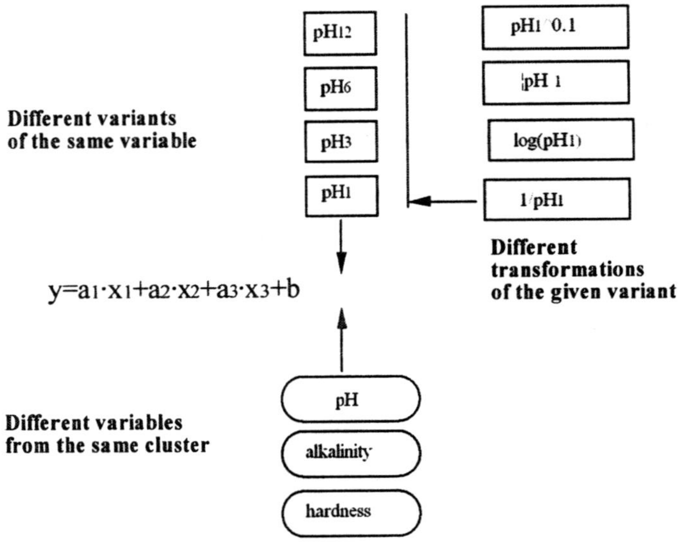

Different **variables** from the same cluster or functional group, e.g.:
• x1 could be a lake morphometric size parameters such as lake area, volume and max. depth
• x2 could be a catchment parameter like Mire% or Forest% belonging to the same cluster.
• x3 could be a water chemical variables like alkalinity, pH, Ca and hardness (= CaMg) belonging to the same cluster.

Fig. 4.18. Illustration of variants, transformations and clusters. In statistical analyses of lake data, it is important to ask questions about the time and area compatibility of the data, i.e., what do the data actually represent?

4.4. Regressions

This section will present many regressions and the main aim is to try to clarify the role of lake morphometry in predicting and understanding the given y-variables. Table 4.8 gives a typical set-up for the following parts of this chapter. It shows first how lake morphometric parameters, then how catchment area parameters, then how lake morphometric plus catchment areas parameters and, finally, how lake morphometric plus catchment area parameters plus water chemical variables can statistically explain variations among lakes in mean annual hardness ($CaMg_{12}$). These calculations use the data in appendix 8.3 and statistical methods discussed in detail by Håkanson and Peters (1995) concerning transformations, time and area compatibility of data, variants (such as mean values from different periods) and procedures of stepwise multiple regressions. Those methods will not be discussed here.

Hardness has a low sedimentation rate (see fig. 4.10), is not retained in lakes very much and cannot be predicted well from lake morphometric data. Table 4.8A shows that 13% ($r^2 = 0.13$) of the variability among these lakes in $CaMg_{12}$ may be statistically explained by variations in relative depth (D_{rel}), the most powerful morphometric predictor. On the other hand, the variation in $CaMg_{12}$ among the lakes may be explained very well by differences in catchment area characteristics. The ladder in table 4.8B shows that $\log(CaMg_{12})$ may be predicted quite well from four catchment parameters. The r^2 value after four steps is 0.67. The regressions are based on

Open land% (OL%), that is, the percentage of the catchment area with open (often cultivated) land; OL% has a positive effect (r^2 = 0.46) on log($CaMg_{12}$); altitude (r^2 increases from 0.46 to 0.60), percentage of outflow areas (OA%, i.e., the upstream lakes and wetlands; r^2 increases to 0.63), and relief [log(RDA); r^2 increases to 0.67]. The relief is defined from the largest height difference in the catchment (RDA = dH/√ADA; RDA = relief of drainage area; dH = the maximum height difference in the drainage area; ADA = area of drainage area in m^2). Note that altitude, outflow areas and relief all have negative effects on $CaMg_{12}$.

Table 4.8C accounts for both morphometric and catchment area parameters. Here the r^2 value reaches 0.71 after six steps. It is important to stress that all parameters are more or less related (see the correlation matrix in table 4.6). This means that the r^2 values are not additive; r^2 = 0.13 from table 4.8A and r^2 = 0.67 from table 4.8B adds up to 0.80, but r^2 from table 4.8C is only 0.71.

In the following discussion, similar r^2 values as from table 4.8A, B and C will be given for many lake variables; and they will also be used in a final analysis of the role of lake morphometry in predicting lake variables.

Table 4.8. Ladder for the mean annual hardness of lake water [y = log($CaMg_{12}$); $CaMg_{12}$ in in meq/l]. The table lists the r^2 values and the regressions at each step. F > 4. Data for 95 lakes (see appendix 8.3).

Step	Parameter	r^2	Model
A. $CaMg_{12}$ vs. morphometric parameters			
1	log(D_{rel})	0.13	$y=-0.642-0.190\cdot x_1$
B. $CaMg_{12}$ vs. catchment parameters			
1	OL%	0.46	$y=-0.748+0.024\cdot x_1$
2	Alt	0.60	$y=-0.556+0.019\cdot x_1-0.001\cdot x_2$
3	log(OA)	0.63	$y=-0.427+0.018\cdot x_1-0.001\cdot x_2-0.115\cdot x_3$
4	log(RDA)	0.67	$y=-0.201+0.016\cdot x_1-0.0009\cdot x_2-0.142\cdot x_3-0.131\cdot x_4$
C. $CaMg_{12}$ vs. morphometric & catchment parameters			
1	OL%	0.46	$y=-0.748+0.024\cdot x_1$
2	Alt	0.60	$y=-0.556+0.019\cdot x_1-0.001\cdot x_2$
3	log(D_{mv})	0.63	$y=-0.485+0.018\cdot x_1-0.001\cdot x_2-0.133\cdot x_3$
4	log(OA)	0.66	$y=-0.352+0.017\cdot x_1-0.0009\cdot x_2-0.136\cdot x_3-0.118\cdot x_4$
5	log(RDA)	0.70	$y=-0.157+0.015\cdot x_1-0.0009\cdot x_2-0.121\cdot x_3-0.142\cdot x_4-0.117\cdot x_5$
6	D_{rel}	0.71	$y=-0.127+0.015\cdot x_1-0.0009\cdot x_2-0.172\cdot x_3-0.146\cdot x_4-0.139\cdot x_5+0.21\cdot x_6$
D. $CaMg_{12}$ vs. morphometric, catchment & water variables			
1	log($cond_{12}$)	0.76	$y=-1.09+0.779\cdot x_1$
2	log(alk_{12}+0.01)	0.90	$y=-0.796+0.745\cdot x_1+0.202\cdot x_2$
3	log(RDA)	0.91	$y=-0.652+0.697\cdot x_1+0.222\cdot x_2-0.061\cdot x_3$

Table 4.8D also includes chemical variables. The r^2 value reaches 0.91 after just three steps, when conductivity (step 1; $r^2 = 0.76$), alkalinity ($r^2 = 0.90$) and relief (RDA) are accounted for in the model. The two first regressions given in table 4.8D simply describe in a quantitative manner how hardness co-vary with conductivity and alkalinity, since they belong to the same cluster of water variables (see table 4.7)

From the foregoing, one can conclude that it is interesting to see how lake variables may be related to:

1. Lake morphometric parameters.
2. Catchment area parameters.
3. Other factors, including climatological variables causing seasonal changes (such as temperature and water discharge), water chemical factors (such as conductivity and alkalinity in table 4.8D) and, evidently also, uncertainties in empirical data.

4.4.1. Predicting lake volume and maximum depth

In practical lake management, it is often necessary to predict in several steps. For example, first limnological state variables, like TP, pH and color, could be predicted from readily available map parameters describing the catchment area and the lake morphometry, then the predicted state variables might be used in other models to predict various measures of the biomass of fish, bottom fauna, etc. (see table 4.1). This approach is called step-by-step predictions. Empirical regression models based on step-by-step predictions are an important and integral part of lake management and research. Note that the concept of step-by-step predictions should not be confused with stepwise multiple regression analysis (Draper and Smith, 1966).

Morphometric parameters are usually important and useful x-variables in lake models. They are generally predictors, not predicted variables. Nevertheless, there are lakes where no depth measurements have been made, so that the mean depth and the volume cannot be known. From two-dimensional topographical maps, one can easily determine the lake area (see chapter 3), the shoreline length, the size of the catchment area and its relief (RDA). But it is not possible to calculate lake volume (V), maximum depth (D_{max}) and mean depth ($D_{mv} = V/A$) unless one has access to a bathymetric map. This section concerns the prediction of such characteristics.

Table 4.9A illustrates a regression between lake area and volume. The transformations that yield the most normal frequency distributions have been used [namely log(A) and log(1000·V), where A is in km^2 and volume in km^3]. Log(1000·V) is used instead of log(V) for convenience. The regression line is: $\log(1000 \cdot V) = 0.605 + 1.142 \cdot \log(A)$. The r^2 value for these 95 lakes is 0.85. This result means that lakes with large areas can be predicted to have large volumes, which has already been shown in fig. 4.17C (see also Straskraba, 1980). Table 4.9A shows that the r^2 value may be further increased by accounting for the relief of the catchment (RDA): The higher the relief, the larger the mean depth and the larger the volume. The r^2 value for the empirical model with these two x-variables is 0.87.

If one has access to empirical data on lake volume (as for these 95 lakes), it is possible to develop models predicting the maximum depth (D_{max} in m), as shown by two regression models in table 4.9B. From empirical data on lake area and volume, one can predict maximum depth; $r^2 = 0.87$. This means that one can also estimate the volume development ($V_d = 3 \cdot D_{mv}/D_{max}$) and hence also the hypsographic form of the lake, as well as important factors in dynamic modeling, such as the ET areas, and hence also where resuspension takes place.

It should also be stressed that step-by-step predictions can also cause large error propagations.

Table 4.9. Regressions for lake morphometric parameters from 95 Swedish lakes and catchments.
A. Predictions of lake volume (V) from lake area (A) and lake area plus catchment relief (RDA).
B. Predictions of maximum depth (D_{max}) using empirical data for lake volume and lake area.

A. Prediction of lake volume (V in km^3); 95 lakes; F > 4		
Model parameters	**r^2**	**Model**
1. A	0.85	$\log(1000 \cdot V)=0.605+1.142 \cdot \log(A)$
2. A, RDA	0.87	$\log(1000 \cdot V)=0.134+1.224 \cdot \log(A)+0.332 \cdot \log(RDA)$
B. Prediction of lake maximum depth (D_{max} in m) using data on lake volume (V); 95 lakes; F > 4		
1. V	0.47	$\log(D_{max})=-0.601+1.499 \cdot (1000 \cdot V)^{0.1}$
2. V, A	0.87	$\log(D_{max})=-4.202+4.558 \cdot (1000 \cdot V)^{0.1}-1.008 \cdot \log(A)$

4.4.2. Predicting key abiotic variables

In this section, the focus will first be set on the three limnological state variables (phosphorus, color and pH) that express important elements of lake structure and function, as shown for TP in table 4.1. A fundamental issue in lake studies presents itself: If there is a change in an abiotic state variable, like TP, color and/or pH, will there also be changes in key functional groups and changes in ecosystem structure? Is it possible to quantify and predict such changes?

Generally, in building empirical/statistical models, it is a requirement that the included x-variables should emanate from different clusters. In complex lake ecosystems, however, where "everything depends on everything else", is not possible find "independent" variables, as this term is often used in more statistical contexts. It is evident that, for example, lake TP, pH and color are determined by independent methods and that they belong to different functional classes and play different roles in aquatic systems, but they are anyhow related. There are many reasons for this. All of them are, at least partly, transported to the lakes by the same tributaries; many internal processes, like wind-induced resuspension and mixing, affect them in similar ways; and they are all transported out of the lake together with all other dissolved and suspended substances.

4.4.2.1. Lake total phosphorus

Total phosphorus (TP) has long been recognized as the nutrient most likely to limit lake primary productivity (Schindler, 1977, 1978; Chapra, 1980; Wetzel, 2001). The concentration of TP in the lake water is a powerful predictor in limnology (see table 4.1). In this section, the regression in table 4.10 shows things the other way around, that is, how mean annual TP concentration (TP_{12}) may be predicted from (A) lake morphometric parameters, (B) catchment parameters and (C) both.

Particulate phosphorus has a relatively low fall velocity (see table 4.5), the particulate fraction is about 0.56 (see fig. 4.19, which gives a regression between TP and PP = particulate P). From these data, one sees that TP is not generally retained very well in lakes; and according to the criteria in the interpretational key, TP cannot be predicted very well from lake morphometric data. One should also note that the highest reference r_r^2 is only 0.92 because the characteristic CV_w is as high as 0.35 for TP (see table 4.2).

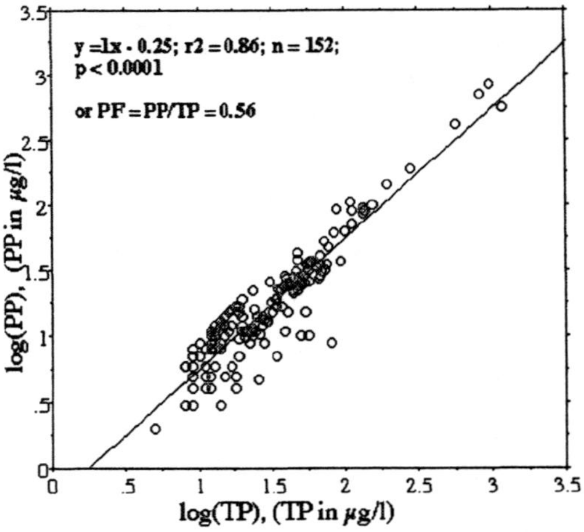

Fig. 4.19. The relationship between the particulate concentration (PP) and the total concentration (TP) for phosphorus based on data from 151 lake samples. From the regression, one can derive that the particulate fraction is 0.56 on average (data from Håkanson and Johansson, 2003).

Table 4.10B shows that 32% ($r^2 = 0.32$) of the variability among these lakes in mean annual TP (TP_{12}) may be statistically explained by variations in the following three catchment area parameters: (1) the percentage of rocks, Rock% ($r^2 = 0.12$), the percentage of open (cultivated) land, OL%, and lake area to catchment area ratio (A/ADA). The r^2 value after three steps is 0.32 (n = 65 lakes) and represents about 35% of the explainable variability (= 0.32/0.92) It should be noted that:

- No single factor gives a very high r^2 value with mean annual TP.
- Rock% appears with a logical minus sign (the more bare rocks in the catchment, the lower the transport of phosphorus from land to water).
- Open land% (OL%) appears with a logical plus sign (the more cultivated land, the higher the transport of phosphorus from land to water).
- The larger the catchment area relative to the lake area, the higher the transport from land to water.

Table 4.10. Ladder for the mean annual concentration of total phosphorus in lake water [$y = \log(TP_{12})$; TP_{12} in µg/l]. The table lists the r^2 values and the models at each step. F > 4. Data from 65 lakes.

Step	Parameter	r^2	Model
A. TP_{12} vs. morphometric parameters			
1	$\log(D_{mv})$	0.15	$y=1.173-0.274 \cdot x_1$
B. TP_{12} vs. catchment parameters			
1	log(1+Rock%)	0.12	$y=1.101-0.110 \cdot x_1$
2	OL%	0.25	$y=1.076-0.129 \cdot x_1+0.015 \cdot x_2$
3	log(A/ADA)	0.32	$y=0.939-0.143 \cdot x_1+0.014 \cdot x_2-0.109 \cdot x_3$
C. TP_{12} vs. morphometric & catchment parameters			
1	$\log(D_{mv})$	0.15	$y=1.173-0.274 \cdot x_1$
2	OL%	0.24	$y=1.140-0.274 \cdot x_1+0.012 \cdot x_2$
3	log(1+Rock%)	0.32	$y=1.171-0.208 \cdot x_1+0.014 \cdot x_2-0.101 \cdot x_3$
4	$T^{0.1}$	0.37	$y=1.541-0.082 \cdot x_1+0.015 \cdot x_2-0.110 \cdot x_3-0.463 \cdot x_4$

Table 4.10C accounts for both morphometric and catchment area parameters. Then the r^2 value is 0.37 after four steps, and the theoretical lake water retention time (T) enters at the last step, which increases r^2 from 0.32 to 0.37.

4.4.2.2. Lake color and pH

4.4.2.2.1. Color

There are many publications on colored substances in aquatic systems (Gorham et al., 1983, 1986; Christman and Gjessing, 1983; Aiken et al., 1985; Hayes et al., 1989). There are several reasons for this interest. Colored substances reduce light transmission through water (Kirk, 1976; Spence, 1982), affect fluxes, bioavailability and ecological effects of nutrients (Peterson, 1991), and of natural elements, like Fe, Al and Pb (Stumm and Morgan, 1981; Baccini et al., 1982) and contaminants (Wershaw et al., 1969; Peterson, 1991), affect plankton metabolism (Jackson and Hecky, 1980; Janus and Vollenweider, 1981; Benner et al., 1986) and may inhibit secondary production (Rasmussen and Kalff, 1987).

Before Rasmussen et al. (1989) presented their quantitative study of the relationship between lake color and characteristics of the drainage area and lake morphometry, only a few predictive models dealt with lake color (Schindler, 1971; Gorham et al., 1986; Engstrom, 1987). Rasmussen et al. (1989) demonstrated that lake color in 337 lakes from Canada and northern U.S.A. was:

- Positively related to the drainage ratio (the ratio between the size of the drainage area and lake area, ADA/A).
- Negatively related to the slope of the drainage area, presumably because drainage areas that are steep or have a high relief generally have thinner organic horizons, fewer bogs and mires and more ground water discharge. Mountain lakes are often exceptionally clear.
- Negatively related to lake mean depth and lake area, suggesting that the turnover of colored substances may also be related to resuspension.

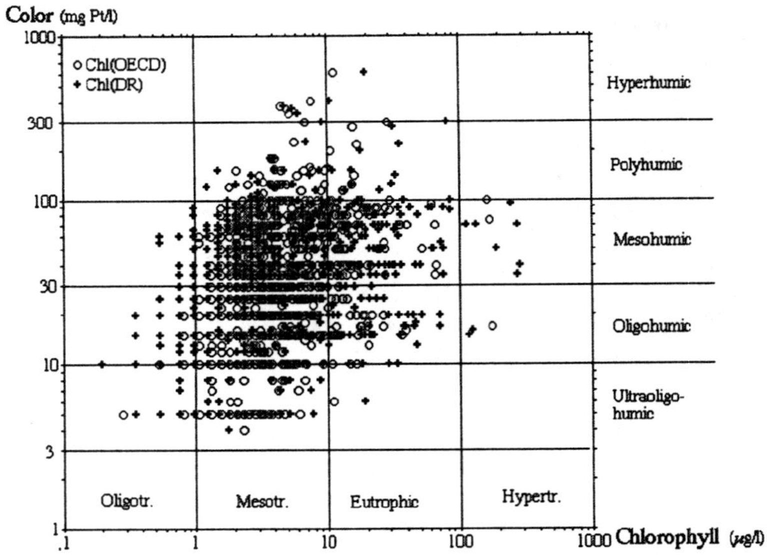

Fig. 4.20. The relationship between lake color (and humic state) on the y-axis and lake chlorophyll (and trophic state) on the x-axis based on data from 936 lakes. Modified from Håkanson and Boulion (2002).

The best general regression model explained 60% of the statistical variance in lake color among these North American lakes.

Fig. 4.20 shows the relationship between lake color on the y-axis, an expression for the humic state of lakes, and chlorophyll concentrations on the x-axis, a measure of the trophic state. The set-up uses (like several similar systems) categories of autotrophy on the x-axis and categories of allotrophy on the y-axis to classify lakes. Using lake color values as a criterion for lake humic state, it is also possible to define a Humic State Index (HSI) on a scale from HSI = 100, hyperhumic, to HSI = 0, ultraoligohumic (color values lower than 3 would be very rare, and if color < 3 mg Pt/l, then HSI = 0) related to the lake characteristic mean values of lake color, accordingly (see Håkanson and Boulion, 2002):

$$HSI = (100/3)\cdot\log(Col - 3) \quad (4.7)$$

Col = 1000 (higher values than 1000 mg Pt/l would be most exceptional) gives HSI = 100; Col = 300 mg Pt/l gives HSI = 82, etc. Note again that if Col < 3, then HSI should be set to 0. The following color values may be used to define the different humic classes: 3, 10, 30, 100 and 300 mg Pt/l for oligo-, meso-, poly- and hyperhumic lakes.

It should also be noted that the secondary production in especially "brown-water" lakes (see Jonsson, 1997) is not only a matter of the supply of phosphorus. Humic lakes can have a substantial production of bacteria and zooplankton driven by allochthonous carbon and nitrogen.

Fig. 4.20 also gives the trophic level on the x-axis using chlorophyll-a (µg/l) as the basic criterion. One can also define a Trophic State Index (TSI), accordingly (see Håkanson and Boulion, 2002):

$$TSI = 25\cdot(\log(Chl) + 1) \quad (4.8)$$

TSI = 100 for Chl = 1000 µg/l (the limit for hypertrophic lakes; Chl = the mean, lake characteristic value for the growing season), TSI = 75 for Chl = 100 µg/l, the limit between eutrophic and hypertrophic lakes, etc. Note that if Chl < 0.1, then TSI should be set to 0. So, TSI is an index that varies from 0 to 100.

The data in fig. 4.20 demonstrate that there exist lakes in most of the "boxes" defined by the humic and trophic categories, except that there are no hypertrophic and ultraoligohumic lakes and no polyhumic/hyperhumic and oligotrophic lakes. That is not to say that such lakes do not exist, only that such lakes are rare.

Table 4.11 gives a ladder for lake color using the database in appendix 8.3. When only lake morphometric parameters are accounted for in regressions against characteristic lake color values (mean values based on data from three years, Co_{l36} in mg Pt/l), then 28% of the variability can be accounted for statistically.

- Theoretical lake water retention time enters at step 1 in the model in table 4.11A. The longer the retention time, the lower the concentration of colored materials in the lake water. This result is logical since more of the colored substances should find the time to be deposited when T is large.
- The mean depth (D_{mv} in m), step 2. The deeper the lake, the lower the values of lake color. In deep lakes, resuspension is smaller and the time for these humic particles to settle is longer

Table 4.11B gives the corresponding results when just catchment area characteristics are included in the regressions.

- The percentage of the drainage area covered in rocks enters at step 1. Where Rock% is high, drainage areas are more infertile so less colored and organic materials are likely to be transported from land to water, thus lowering the values of lake color. This reasoning seems logical, but it is interesting to note that this factor was more important than Mire%, Lake% and Forest%.
- In agreement with results presented by Rasmussen et al. (1989), the drainage ratio enters at step 2 in a logical way and increases r^2 from 0.24 to 0.45.
- Lake% enters at the third step (r^2 = 0.58). The higher the Lake%, the more colored substances are trapped upstream and the lower the color values in a given lake.
- The more mires, the higher the color values in the lake (r^2 = 0.63). This relationship is logical and expected.
- Also the percentage of till in the catchment increases lake color values (r^2 = 0.66).
- In the final step in the model based on catchment parameters, ones find altitude. The higher the altitude, the lower the lake color values (r^2 = 0.69).

All other parameters describing lake morphometry and drainage area characteristics are weakly correlated with lake color. Color values may be statistically explained very well by differences in catchment area characteristics among the lakes.

Table 4.11C accounts for both morphometric and catchment area parameters. Here, the r^2 value reaches 0.68 after six steps, which is lower than 0.69 from table 4.11B. This result is by chance.

Table 4.11. Ladder for the mean lake color based on data from 36 months [$y = \log(Col_{36})$; Col_{36} in mg Pt/l]. The table lists the r^2 values and the models at each step. F > 4. Data from 91 lakes.

Step	Parameter	r^2	Model
A. Col_{36} vs. morphometric parameters			
1	$T^{0.1}$	0.24	$y=2.913-1.024\cdot x_1$
2	$\log(D_{mv})$	0.28	$y=2.721-0.695\cdot x_1-0.231\cdot x_2$
B. Col_{36} vs. catchment parameters			
1	log(1+Rock%)	0.24	$y=2.089-0.213\cdot x_1$
2	$(A/ADA)^{0.3}$	0.45	$y=2.504-0.262\cdot x_1-0.902\cdot x_2$
3	log(1+Lake%)	0.58	$y=2.747-0.250\cdot x_1-1.2530\cdot x_2-0.269\cdot x_3$
4	Mire%	0.63	$y=2.617-0.217\cdot x_1-1.207\cdot x_2-0.283\cdot x_3+0.006\cdot x_4$
5	Till%	0.66	$y=2.360-0.149\cdot x_1-1.248\cdot x_2-0.260\cdot x_3+0.009\cdot x_4+0.003\cdot x_5$
6	Alt	0.69	$y=2.269-0.121\cdot x_1-1.202\cdot x_2-0.245\cdot x_3+0.012\cdot x_4+0.005\cdot x_5-0.0007\cdot x_6$
C. Col_{36} vs. morphometric & catchment parameters			
1	log(1+Rock%)	0.24	$y=2.089-0.213\cdot x_1$
2	$T^{0.1}$	0.49	$y=3.075-0.215\cdot x_1-0.215\cdot x_2$
3	log(1+Lake%)	0.56	$y=3.309-0.193\cdot x_1-1.230\cdot x_2-0.195\cdot x_3$
4	log(OA)	0.65	$y=2.990-0.889\cdot x_1-1.194\cdot x_2-0.280\cdot x_3+0.255\cdot x_4$
5	Till%	0.66	$y=2.734-0.103\cdot x_1-1.199\cdot x_2-0.271\cdot x_3+0.313\cdot x_4+0.02\cdot x_5$
6	$(A/ADA)^{0.3}$	0.68	$y=2.565-0.127\cdot x_1-0.710\cdot x_2-0.301\cdot x_3+0.286\cdot x_4+0.0025\cdot x_5-0.581\cdot x_6$

4.4.2.2.2. pH

Lake pH is also a very important lake variable. Fig. 4.21 shows that many animals accustomed to a circum-neutral pH (pH ≈ 7) cannot reproduce and survive in acidified lakes. Some, like crayfish, crustaceans and snails, are very sensitive to changes in pH, whereas other animals, like perch and pike, are less sensitive. The literature on anthropogenic acidification of land and water, its ecological damage and its economic consequences is extensive (Likens et al., 1979; Ambio, 1976; Overrein et al., 1980; Merilehto et al., 1988).

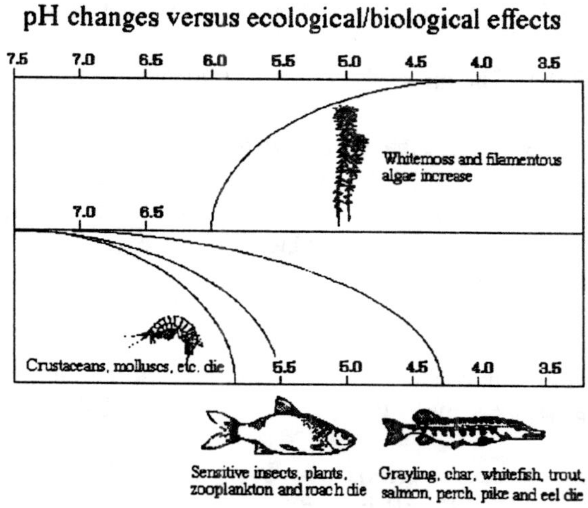

Fig. 4.21. Illustration of how changes in lake pH influence different organisms. The figure shows examples of key functional groups and target organisms for acidification. Crustaceans react rapidly to changes in pH, whereas certain fish such as brook trout and eels do not die until acidification is far advanced. White moss (e.g., *Spagnum*) and filamentous algae should not be found in these lakes under normal conditions. So, the abundance of such species also indicates ecological effects of acidification (modified from Håkanson and Boulion, 2002).

Ivanova (1997) reports that zooplankton exist at pH from 3.5 to 10.5 and that reductions in species number can be observed in lakes with pH $\leq$5.5. It seems that pH of about 5.5 may be boundary condition for the structure of the zooplankton community. It is important to stress that in acidic lakes, the zooplankton biomass generally exceeds the phytoplankton biomass during the growing season (Ivanova et al., 1993; Krylov et al., 1997). As a result, the consumption by herbivorous zooplankton exceeds the phytoplankton production. One can then safely assume that the main primary producers in acidic lakes are water moss covering the bottom (see fig. 4.21) and phytobenthos growing on moss.

The ladder for mean annual lake pH (pH_{12}) is given in table 4.12. From this table, one can see that:

- pH_{12} may not be predicted very well from lake morphometric parameters. There is a very weak relationship between lake area and pH (r^2 = 0.07). The causal reasons for this low r^2 value are not clear, but a larger area would imply a larger bioproduction and that would increase lake pH. The main reason for the low r^2 value has been discussed and relates to the ratio between CV_a and CV_w, which is close to 1 for pH (see fig. 4.9). Lake pH is known to be difficult to predict (see Håkanson and Peters, 1995) since it depends on acid rain, transport of substances from land influencing the H^+ activity and substances buffering the H^+ activity, on seasonal variations in lake bioproduction, etc.

Table 4.12B shows the predictive power obtained using catchment area parameters. The r^2 value after seven steps is 0.46, and that model is based on:

- Open land% (OL%). This positive correlation is logical since more open (cultivated) land implies more fine sediments, more ions, higher buffering capacity, more agricultural fertilizers (with nutrients), and thus higher nutrient loading to the lakes, a higher production and a higher lake pH.
- Mire%. As expected, the correlation is negative because more mires produce more humic substances, which lower lake pH.
- Relief (RDA). The statistically significant effect of relief is difficult to mechanistically

explain, but the statistical correlation is nevertheless positively significant for these lakes.

- Catchment area (ADA). The larger the catchment in absolute terms, the higher the lake pH.
- ADA/A. The larger the catchment area relative to the lake area, the higher the pH.
- Altitude. There is a negative relationship between lake pH and altitude (lakes at higher altitudes often have lower pH). This fact may be an effect of temperature and lower production related to lower temperatures at higher altitudes.
- Rock%. There is a logical negative relationship between rocks in the catchment and low lake pH.

Table 4.12C shows the regressions obtained using both lake morphometry and catchment area parameters. The r^2 value after six steps is 0.46, and that model is based on the parameters just mentioned.

It is interesting to note that the characteristic within-lake variability (CV_w) for pH is small (0.05; see table 4.2). This small CV_w for pH indicates that it would be possible to predict lake pH well, but these models do not do that. One should then note that pH by definition is a logarithm. Basically, the H^+ activity is likely to vary much like alkalinity with a CV_w value of 0.35 and a corresponding r_r^2 value of 0.92.

Table 4.12. Ladder for the mean annual pH of lake water ($y = pH_{12}$). The table lists the r^2 values and the models at each step. $F > 4$. Data from 93 lakes.

Step	Parameter	r^2	Model
A. pH_{12} vs. morphometric parameters			
1	log(A)	0.07	$y=5.968+0.208 \cdot x_1$
B. pH_{12} vs. catchment parameters			
1	log(1+OL%)	0.13	$y=5.726+0.376 \cdot x_1$
2	Mire%	0.21	$y=5.976+0.303 \cdot x_1-0.015 \cdot x_2$
3	log(RDA)	0.26	$y=5.409+0.385 \cdot x_1-0.013 \cdot x_2+0.330 \cdot x_3$
4	$\sqrt{ADA}$	0.32	$y=5.081+0.364 \cdot x_1-0.015 \cdot x_2+0.431 \cdot x_3+0.062 \cdot x_4$
5	log(ADA/A)	0.39	$y=5.180+0.338 \cdot x_1-0.016 \cdot x_2+0.564 \cdot x_3+0.101 \cdot x_4-0.309 \cdot x_5$
6	$\sqrt{Alt}$	0.44	$y=5.598+0.269 \cdot x_1-0.014 \cdot x_2+0.626 \cdot x_3+0.104 \cdot x_4-0.367 \cdot x_5-0.035 \cdot x_6$
7	log(1+Rock%)	0.46	$y=5.754+0.241 \cdot x_1-0.016 \cdot x_2+0.666 \cdot x_3+0.110 \cdot x_4-0.351 \cdot x_5-0.044 \cdot x_6-0.153 \cdot x_7$
C. pH_{12} vs. morphometric & catchment parameters			
1	log(1+OL%)	0.13	$y=5.726+0.376 \cdot x_1$
2	Mire%	0.21	$y=5.976+0.303 \cdot x_1-0.015 \cdot x_2$
3	$\sqrt{A}$	0.28	$y=5.847+0.242 \cdot x_1-0.019 \cdot x_2+0.269 \cdot x_3$
4	log(RDA)	0.38	$y=4.902+0.347 \cdot x_1-0.017 \cdot x_2+0.378 \cdot x_3+0.520 \cdot x_4$
5	$\sqrt{Alt}$	0.42	$y=5.223+0.283 \cdot x_1-0.014 \cdot x_2+0.402 \cdot x_3+0.572 \cdot x_4-0.033 \cdot x_5$
6	log(1+Rock%)	0.46	$y=5.443+0.244 \cdot x_1-0.017 \cdot x_2+0.440 \cdot x_3+0.635 \cdot x_4-0.045 \cdot x_5-0.188 \cdot x_6$

4.4.2.3. Water chemistry (alkalinity, calcium, iron and conductivity)

These standard water variables play many different roles in lakes (see Wetzel, 2001). Alkalinity, calcium, hardness (CaMg), conductivity and pH form a cluster of variables showing high internal correlations. Fe is more related to color (see table 4.7). Can these variables be related to lake morphometry?

Table 4.13 shows the results for alkalinity. One can note that:

- Alkalinity cannot be predicted by lake morphometric variables. The reason is that mass-balances (see table 4.5) show that inflow equals outflow for alkalinity and that alkalinity is a conservative variable. So, this is a logical result.
- The ladder for alkalinity versus catchment parameters (table 4.13B) shows that the r^2 value after four steps is just 0.34. So, $\log(alk_{12}+0.01)$, which is the transformation yielding the most normal frequency distribution, is not predicted very well from standard catchment area characteristics. The model is based on the same parameters as the model for pH_{12}, and the same "explanations" may be used for the significant positive effects of Open land% (OL%), relief (RDA) and catchment area (ADA), and the negative relationship with altitude (Alt).

Table 4.14 gives the results for calcium (Ca_{36}). Logically, these results are very similar to the results for hardness (CaMg; see table 4.8). One can note that:

- $\log(Ca_{36})$ may be predicted quite well from catchment parameters (the r^2 value after three steps is 0.55) but not from lake morphometric parameters (r^2 = 0.08) because calcium is basically very conservative (see table 4.5).
- If morphometric and catchment parameters are pooled, the r^2 value is 0.62. Altitude is the most important x-variable (the higher the altitude, the lower lake calcium concentrations). There are positive correlations to open land% (OL%), the dynamic ratio (DR) and the relative depth (D_{rel}), but these are weaker than the correlation to the shoreline development (L_d).

Iron has been determined as "acid-soluble" Fe and analyzed mainly to study the amounts of suspended humus and iron oxides and hydroxides. The concentration of iron belongs to the same cluster as lake color. Highly significant relationships between Fe and color have been demonstrated in previous works (see Wetzel, 2001). The "ladders" for the 3-year mean values of lake Fe concentrations (Fe_{36}), as predicted from map characteristics, are given in table 4.15. Because of the high correlation between Fe and color, these results are similar to the results shown in table 4.11 for color. One can note that:

- Mire% (r^2 = 0.32) and Till% appear with positive correlations towards Fe_{36}. All other parameters [mean depth, outflow areas (= OA), altitude, mean water discharge (Q_{mv}), maximum depth, Rock% and the A/ADA ratio] appear with negative signs. After six steps r^2 is 0.71 in table 4.15C.
- The highest reference r_r^2 (see table 4.2) for the lake Fe concentration is 0.96. Thus, all other factors that could potentially (statistically) affect Fe give an unexplained residual of about 25% (0.96-0.71).
- It is likely that Fe, like color, is rather conservative (see table 4.5), which explains the low correlation towards lake morphometric parameters.

Table 4.13. Ladder for the mean annual alkalinity of lake water [$y = \log(0.01+alk_{12})$; alk_{12} in meq/l]. The table lists the r^2 values and the models at each step. $F > 4$. Data from 93 lakes.

Step	Parameter	r^2	Model
A. alk_{12} vs. morphometric parameters; no significant relationships			
B. alk_{12} vs. catchment parameters			
1	log(1+OL%)	0.11	$y=-1.484+0.268 \cdot x_1$
2	log(RDA)	0.25	$y=-2.220+0.367 \cdot x_1+0.459 \cdot x_2$
3	Alt	0.30	$y=-2.027+0.297 \cdot x_1+0.486 \cdot x_2-0.001 \cdot x_3$
4	√ADA	0.34	$y=-2.259+0.296 \cdot x_1+0.551 \cdot x_2-0.001 \cdot x_3+0.036 \cdot x_4$
C. alk_{12} vs. morphometric & catchment parameters; same as B.			

Table 4.14. Ladder for the calcium concentration in lake water based on data from 36 months [$y = \log(Ca_{36})$; Ca_{36} in meq/l]. The table lists the r^2 values and the models at each step. $F > 4$. Data for 49 lakes.

Step	Parameter	r^2	Model
A. Ca_{36} vs. morphometric parameters			
1	$\log(L_d)$	0.08	$y=-0.881+0.373 \cdot x_1$
B. Ca_{12} vs. catchment parameters			
1	Alt	0.39	$y=-0.465-0.002 \cdot x_1$
2	√OL%	0.51	$y=-0.591-0.001 \cdot x_1+0.060 \cdot x_2$
3	$RDA^{0.1}$	0.55	$y=0.042-0.001 \cdot x_1+0.061 \cdot x_2-0.454 \cdot x_3$
C. Ca_{12} vs. morphometric & catchment parameters			
1	Alt	0.39	$y=-0.465-0.002 \cdot x_1$
2	√OL%	0.51	$y=-0.591-0.001 \cdot x_1+0.060 \cdot x_2$
3	$DR^{0.1}$	0.55	$y=-1.233-0.001 \cdot x_1+0.047 \cdot x_2+0.822 \cdot x_3$
3	$\sqrt{D_{rel}}$	0.62	$y=-3.211-0.002 \cdot x_1+0.039 \cdot x_2+2.862 \cdot x_3+0.258 \cdot x_4$

The results for conductivity [$\log(cond_{12})$] are shown in table 4.16.

- Like hardness and Ca, $cond_{12}$ may be predicted very well from standard catchment parameters. The r^2 value after five steps is 0.69. The model is based on variables similar to those in the models for Ca and CaMg.
- Conductively is a conservative variable, just like calcium, sodium and magnesium, so lake morphometry does not predict conductivity well ($r^2 = 0.15$ from the relative depth).
- The pooled parameters give an r^2 of 0.72, which is high (see table 7.16C).

It is interesting to note that the interpretational key related to the factors influencing the variability within and among lakes and the mass-balance considerations provides very useful information in understanding the given results for all these water chemical variables.

Table 4.15. Ladder for the mean concentration of iron in lake water based on data from 36 months [$y = \log(Fe_{36})$; Fe_{36} in µg/l]. The table lists the r^2 values and the models at each step. $F > 4$. Data from 41 lakes

Step	Parameter	r^2	Model
A. Fe_{36} vs. morphometric parameters			
1	$\log(D_{mv})$	0.10	$y=2.868-0.387\cdot x_1$
B. Fe_{36} vs. catchment parameters			
1	$Mire\%^{0.7}$	0.32	$y=2.290+0.213\cdot x_1$
2	OA%	0.44	$y=2.204+0.164\cdot x_1-0.035\cdot x_2$
3	$(A/ADA)^{0.3}$	0.51	$y=2.436+0.188\cdot x_1-0.043\cdot x_2-0.578\cdot x_3$
4	Till%	0.56	$y=2.363+0.156\cdot x_1-0.035\cdot x_2-0.792\cdot x_3+0.003\cdot x_4$
5	log(Alt)	0.61	$y=3.098+0.184\cdot x_1-0.037\cdot x_2-0.787\cdot x_3+0.007\cdot x_4-0.511\cdot x_5$
6	$Q_{mv}^{0.1}$	0.69	$y=3.893+0.163\cdot x_1-0.028\cdot x_2-1.117\cdot x_3+0.009\cdot x_4-0.554\cdot x_5-0.846\cdot x_6$
C. Fe_{36} vs. morphometric & catchment parameters			
1	$Mire\%^{0.7}$	0.32	$y=2.290+0.213\cdot x_1$
1	$\log(D_{max})$	0.48	$y=2.620+0.067\cdot x_1-0.389\cdot x_2$
3	OA%	0.57	$y=2.509+0.155\cdot x_1-0.343\cdot x_2-0.029\cdot x_3$
4	$(A/ADA)^{0.3}$	0.62	$y=2.707+0.176\cdot x_1-0.328\cdot x_2-0.037\cdot x_3-0.526\cdot x_4$
5	log(1+Rock%)	0.67	$y=2.975+0.154\cdot x_1-0.331\cdot x_2-0.033\cdot x_3-0.717\cdot x_4-0.130\cdot x_5$
6	$\log(L_d)$	0.71	$y=2.934+0.163\cdot x_1-0.363\cdot x_2-0.037\cdot x_3-0.800\cdot x_4-0.149\cdot x_5+0.349\cdot x_6$

Table 4.16. Ladder for the mean annual conductivity of lake water [$y = \log(cond_{12})$; $cond_{12}$ in mS/m]. The table lists the r^2 values and the models at each step. $F > 4$. Data from 94 lakes.

Step	Parameter	r^2	Model
A. $cond_{12}$ vs. morphometric parameters			
1	$\log(D_{rel})$	0.15	$y=0.572-0.225\cdot x_1$
B. $cond_{12}$ vs. catchment parameters			
1	OL%	0.49	$y=0.452+0.028\cdot x_1$
2	log(RDA)	0.59	$y=0.811+0.024\cdot x_1-0.228\cdot x_2$
3	log(Q)	0.64	$y=0.783+0.013\cdot x_1-0.276\cdot x_2-0.099\cdot x_3$
4	Mire%	0.68	$y=0.901+0.021\cdot x_1-0.296\cdot x_2-0.986\cdot x_3-0.005\cdot x_4$
5	log(1+Rock%)	0.69	$y=0.855+0.021\cdot x_1-0.307\cdot x_2-0.097\cdot x_3-0.005\cdot x_4+0.51\cdot x_5$
C. $cond_{12}$ vs. morphometric & catchment parameters			
1	OL%	0.49	$y=0.452+0.028\cdot x_1$
2	log(RDA)	0.59	$y=0.811+0.024\cdot x_1-0.228\cdot x_2$
3	log(V)	0.64	$y=-0.667+0.023\cdot x_1-0.277\cdot x_2-0.077\cdot x_3$
4	log(OA)	0.69	$y=-0.916+0.021\cdot x_1-0.303\cdot x_2-0.073\cdot x_3-0.164\cdot x_4$
5	Alt	0.72	$y=-0.989+0.018\cdot x_1-0.283\cdot x_2-0.067\cdot x_3-0.148\cdot x_4-0.0006\cdot x_5$

4.4.2.4. Suspended particulate matter

Suspended particulate matter (SPM = seston) plays many important roles in lake systems. For example, SPM regulates the two major transport routes, the dissolved pelagic route and the particulate sedimentation and benthic route, of all types of materials and contaminants. The fraction of any given substance in dissolved phase (in mg X/l) and the fraction in particulate phase (in mg X/mg dw; dw = dry weight) are directly related to the concentration of SPM (in mg dw/l) in the water. The carbon content of SPM is crucial at lower trophic levels as a source of energy for bacteria, phytoplankton and zooplankton (Jørgensen and Johnsen, 1989). SPM is also important at higher hierarchical levels, where SPM influences many metabolic properties (Khailov, 1974; Ostapenia, 1989). SPM is also directly related to many variables of general use in water management as indicators of water clarity (such as Secchi depth, color and the depth of the photic zone; Håkanson and Boulion, 2002). These variables are important regulators of primary production and hence also of the secondary production, for example, of zooplankton and fish. The effects of SPM on recycling processes of organic matter, nutrients and pollutants determine the ecological significance of SPM in any given aquatic system. Understanding the mechanisms that control the distribution of SPM in lakes is an issue of both theoretical and applied concern.

Many contaminants (like radionuclides, metals and nutrients) may be removed from the water column due to their sedimentation with SPM and burial in the bottom sediments (IAEA, 2000). The limit between the particulate and the dissolve phase is often determined by means of filtration, using a pore size of 0.45 µm. This approach is simply an operational one, and many colloidal particles will pass such filters (Boulion, 1994).

Fig. 4.22 classifies SPM according to origin. Many factors are known to influence SPM in aquatic systems (Ostapenia et al., 1985; Preisendorfer, 1986; Boulion, 1994, 1997). The most important factors are:

1. Autochthonous production (the amount of plankton, faeces, etc. in the water and more plankton means a higher SPM).
2. Allochthonous materials, such as the amount of colored matter (humic and fulvic substances).
3. The amount of resuspended material.

The total amount of SPM in water is generally a complex mix of substances of different origins with different properties (size, form, density, specific surface area, capacity to bind pollutants, etc.). SPM may be divided into an organic fraction (POM) and an inorganic one (PIM, particulate inorganic materials). Total organic matter (TOM) is generally divided into particulate (POM) and dissolved (DOM) fractions. Normally, POM is about 20% of TOM, but this proportion certainly varies among and within systems (Ostapenia, 1987, 1989; Velimorov, 1991; Boulion, 1994). Normally, about 4% of POM is living matter and the rest is dead organic matter (detritus). About 80% of TOM is generally in the dissolved phase, and of this percentage about 70% is conservative in the sense that it does not change due to chemical and biological reactions in the water mass.

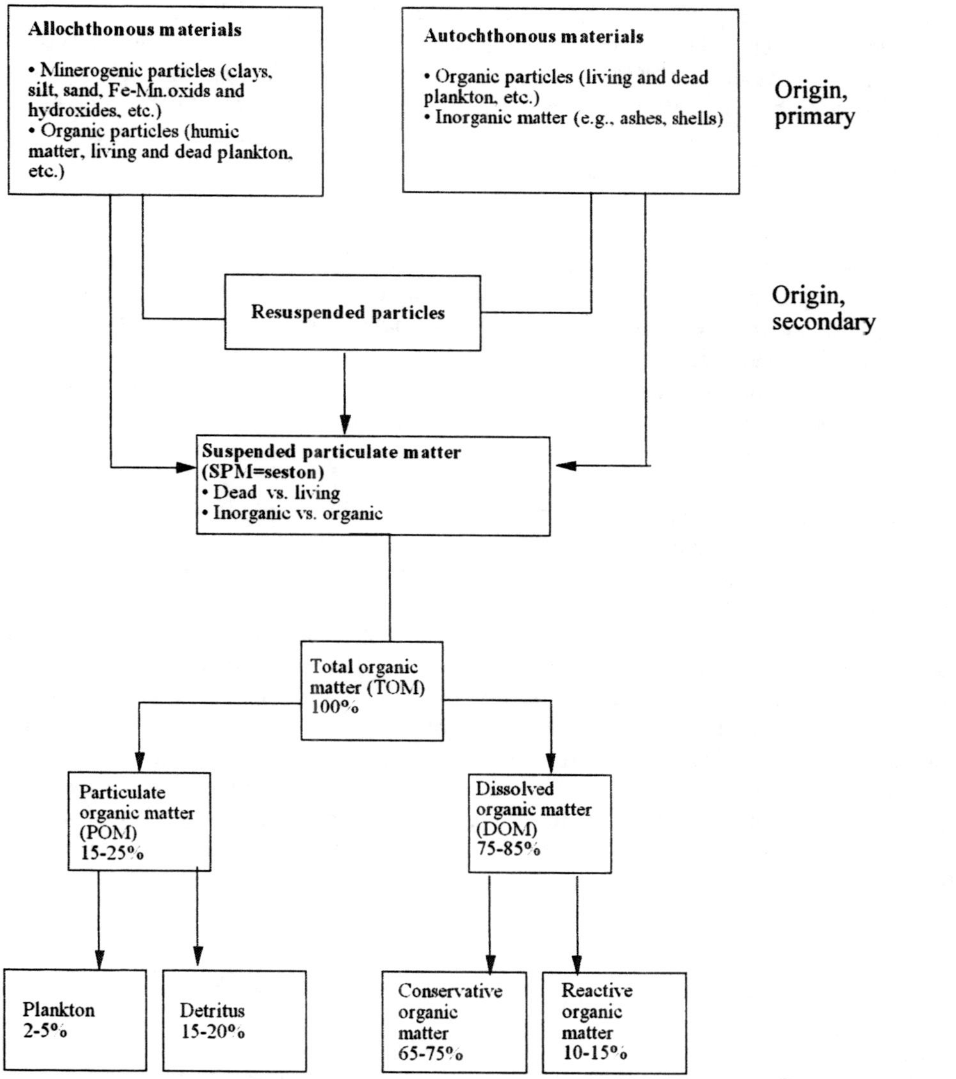

Fig. 4.22. Classification scheme and nomenclature for particles in lake water (see also Dubko, 1985; Ostapenia, 1985).

The particles settling on the bed can be consumed by zoobenthos, and zoobenthos by fish, so that these particles can be transported up to the water column again in this manner. These particles can also be transported by migrating zooplankton between the sediments and the water. Organic particles will also be decomposed by bacteria and then the newly formed dissolved phase can migrate by diffusive processes. Bioturbation can also contribute to the transport of settled particles both back to the water and downward to underlying sediments.

Fig. 4.23 illustrates typical variations in SPM in lakes. It is well known for lakes (Håkanson and Jansson, 1983) that sedimentation increases from about zero at the water depth separating areas of fine sediment transport and accumulation, to maximum values at the deepest part of the basin; and the concentration of SPM increases due to this fact because the volume (defined for a certain water depth) decreases with increasing water depths, a phenomenon often referred to as "sediment focusing" (see fig. 4.23, left). Fig. 4.23, right illustrates the concept "coastal focusing" for coastal areas in large lakes in analogy with sediment focusing. SPM can be expected to appear with relatively high values in coastal areas where the river input of SPM is significant and accentuated by the coastal current and small water depths. Hence, the volume is small and resuspension high. The pattern shown in fig. 4.23, right, is a typical pattern for SPM in coastal areas of large lakes, but this pattern may be modified by many processes, see fig. 4.24, which are all more or less related to morphometry.

- A high river inflow will increase SPM.
- Autochthonous production will increase SPM. The bioproduction is high in areas where phytoplankton, benthic algae and macrophytes contribute to the primary production; in open water areas, only phytoplankton dominate the primary production.
- Sedimentation is a key process to reduce suspended particles in the water mass.
- Resuspension of fine sediments from areas of erosion and transport, that is, advective transport of matter deposited under calm conditions and then resuspended by wind-wave activity or slope processes.

Mean values of suspended particulate matter (SPM in mg/l) may be predicted from a regression model (see table 4.17). This empirical model, which was derived from a broad set of data (see Lindström et al., 1999), gave an r^2 of 0.87 after three steps. The most important model parameter (explaining 74% of the SPM variability among the lakes) was the TP concentration, indicating that the autochthonous production is the most important process determining SPM in these lakes. Lake pH (a collective variable for aggregation and allochthonous and autochthonous processes influencing SPM) and the dynamic ratio (resuspension; $DR = \sqrt{A}/D_{mv}$, where A is in km^2 and mean depth, D_{mv}, is in m) both add significantly to the predictive power of the model. It is interesting to note that a lake morphometric parameter adds significant predictive power to this model. This outcome is also logical since SPM by definition will settle out in lakes, so morphometry should be important, according to the interpretational key. It is also interesting to note that the highest reference r^2 (r_r^2) for SPM is 0.97. So, the unexplained residual term is about 10% (0.97-0.87).

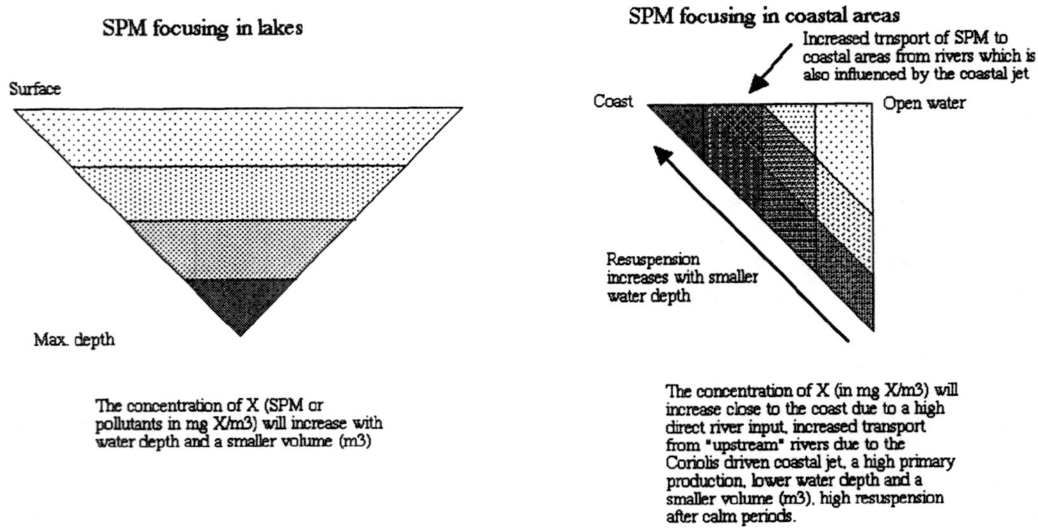

Fig. 4.23. Illustration of sediment focusing in small and "normal" lakes (left) and in coastal areas in large lakes (right).

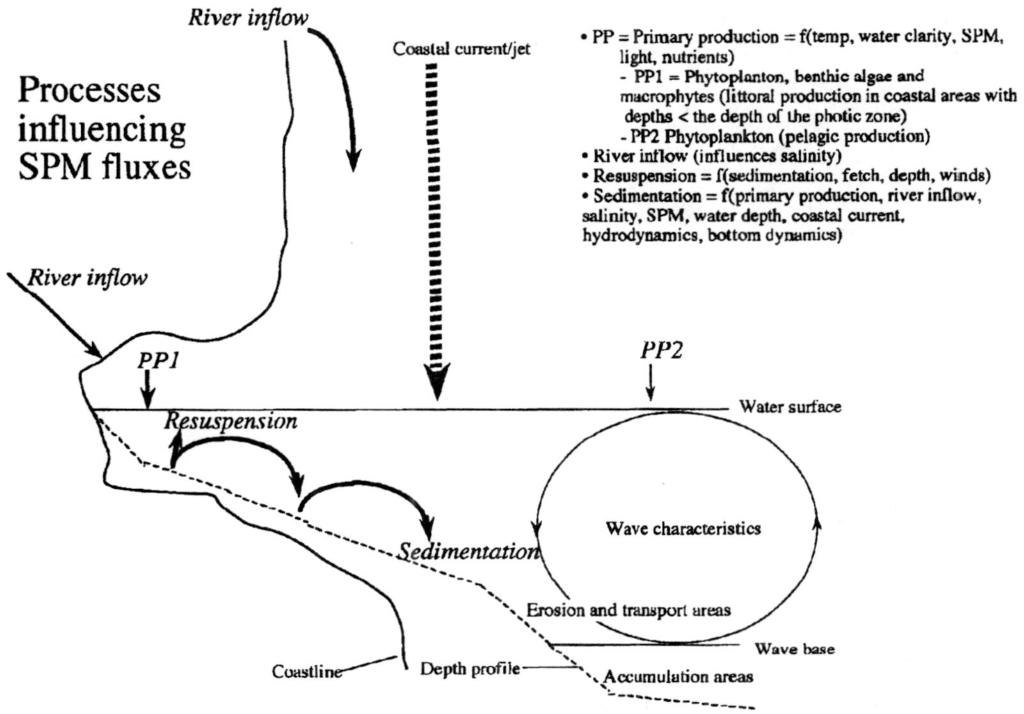

Fig. 4.24. Processes influencing SPM fluxes. Note that many of the processes and factors influencing SPM are interrelated in a complex manner. There is also often a very dynamic exchange of water, SPM and pollutants between the open water areas and coastal areas.

Table 4.17. Ladder for suspended particulate matter (SPM in mg/l) from stepwise multiple regressions based on data from 26 lakes (from Lindström et al., 1999). Dynamic ratio, DR = $\sqrt{A}/D_{mv}$; A = lake area in km^2, D_{mv} = mean depth in m. TP is the mean, characteristic TP concentration in µg/l

Step	r^2	Model variable	Model
1	0.74	TP	log(SPM) = 1.56·log(TP) - 1.69
2	0.83	pH	log(SPM) = 1.47·log(TP) + 0.18·pH - 2.85
3	0.87	DR	log(SPM) = 1.148·log(TP) + 0.137·pH + 0.286·log(DR) - 1.985

The ranges of the model parameters for this model are:

Parameter	Unit	Minimum value	Maximum value
TP	[µg/l]	5	60
pH	[-]	5.1	8.5
DR	[-]	0.07	7.9

The SPM model is graphically shown in fig. 4.25. This figure stresses the role of the dynamic ratio, the lake morphometry, in predicting and understanding SPM variations among lakes. One can note that (for a lake with TP = 20 µg/l), when mean pH values vary between 5 and 9, variations in DR will cause SPM values to vary from about 1 to 8 mg/l.

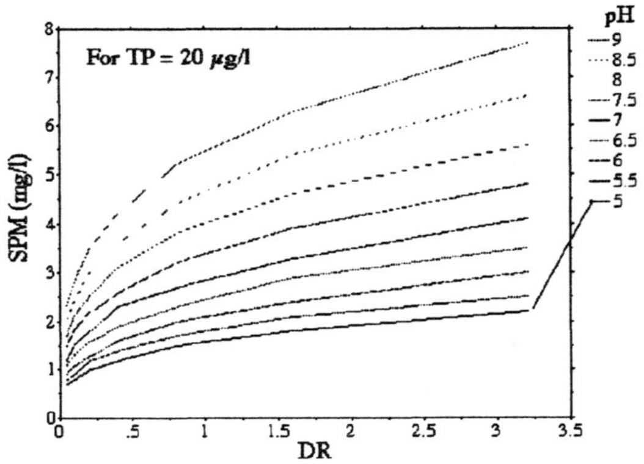

Fig. 4.25. Nomogram illustrating how SPM is related to the dynamic ratio (DR) and pH. In this plot, the lake TP concentration is set to 20 µg/l.

4.4.2.5. Secchi depth

To most people, clear waters with large Secchi depths seem more attractive than turbid waters. Many factors are known to influence the Secchi depth (Vollenweider, 1958, 1960; Carlson, 1977, 1980; Ostapenia et al., 1985; Preisendorfer, 1986). Like SPM, Secchi depth depends on, (1) autochthonous production (the amount of plankton, detritus, etc. in the water and more plankton, etc. mean a lower Secchi depth); (2) allochthonous materials, such as the amount of colored matter in the lake (such as humic, fulvic and minerogenic substances from tributaries); and (3) the amount of resuspended material (materials resuspended from the lake bed via wind/wave activity, slope processes, etc.). These factors are not independent: High sedimentation leads to high amounts of resuspendable materials; high resuspension leads to high internal loading of nutrients and increased production; a high amount of colored substances means a smaller photic zone and a lower production; a high input of colored substances and a high production would mean a high sedimentation, etc. Wallin et al. (1992) showed that the Secchi depth should be much greater than that observed if only plankton cells were responsible for the light extinction. This observation means that particles other than plankton cells are perhaps the most important factors for determining the Secchi depth. It was also concluded that the empirical relationship between Secchi depth and chlorophyll-a largely depends on the chlorophyll-a concentration co-varying with the total amount of suspended particles. This correlation has been discussed also in other studies (Kiefer and Austin, 1974; Tilzer, 1988).

The maximum depth of the photic zone is generally set to about two times the Secchi depth, and the effective (or mean) depth of the photic zone is often set equal to the Secchi depth (see Håkanson and Peters, 1995). Generally, light extinction with water depth is given by the following equation:

$$I(x) = I_0 \cdot e^{-kx} \qquad (4.9)$$

Where

$I(x)$ = a measure of the amount of light at water depth x;
I_0 = the amount of light at the initial reference water depth (x = 0);
k = the light extinction coefficient.

Fig. 4.26 shows the very strong and logical relationship between Secchi depth and the concentration of suspended particulate matter (SPM in mg/l) based on 573 samples from 23 lakes that cover a wide limnological domain from oligotrophic to highly eutrophic conditions.

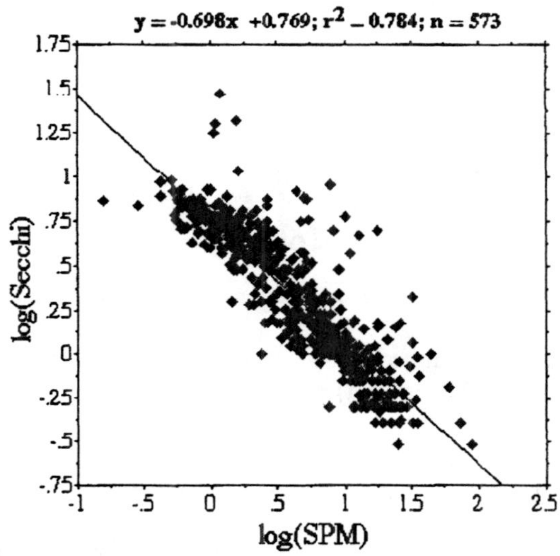

Fig. 4.26. The relationship between Secchi depth (in m) and the concentration of suspended particulate matter (SPM in mg/l) based on 573 data from 23 lakes covering a wide limnological domain (from highly eutrophic to oligotrophic conditions). The figure gives the regression line, r^2 = coefficient of determination, n = number of lakes and p = the statistical certainty (for more information, see Håkanson and Boulion, 2002).

Several studies have quantified and ranked variables of significance to predict how mean values of Secchi depth vary among lakes (see Håkanson and Lindström, 1997; Nürnberg and Shaw, 1998). The most important "water" variables were: Lake color (expressing allogenic input of different types of humic materials), TP concentration and lake temperature (measures of production of autogenic materials). The most important "geological" parameters were: the mean depth (linked to resuspension) and the ratio between the drainage area and lake area (expressing the linkage between catchment and lake).

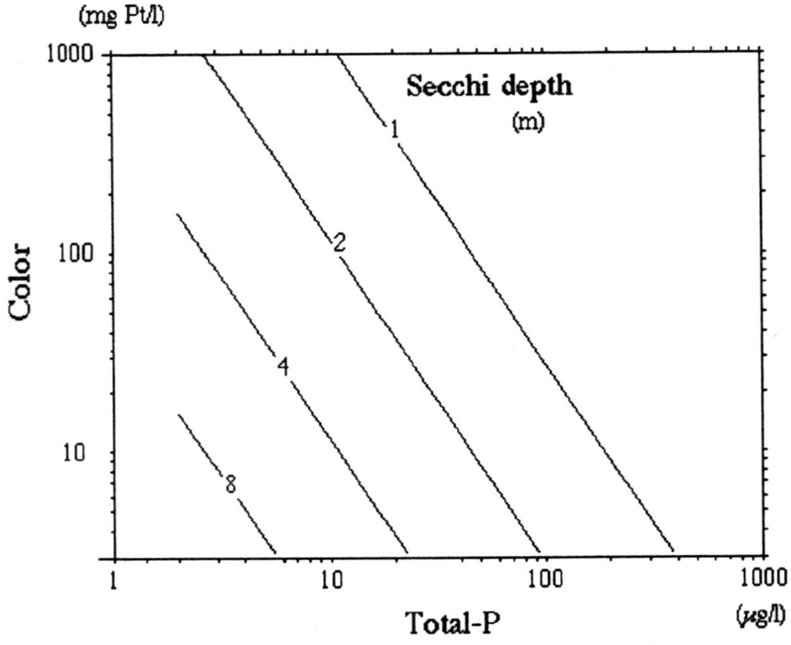

Fig. 4.27. The relationship between lake color, TP concentrations and Secchi depth (from Håkanson and Boulion, 2002).

An empirical model based on data from 935 lakes covering a very wide range gives Secchi depth (Sec in m) as a function of TP (in µg/l) and lake color (Col in mg Pt/l). It is also illustrated in fig. 4.27 and gives:

$$\log(Sec) = 1.41 - 0.49 \cdot \log(TP) - 0.30 \cdot \log(Col) \quad (4.10)$$

Since color is an indicator of lake humic status and lake TP an indicator of lake trophic status (see fig. 4.17), Secchi depth is evidently closely related to these two limnological state variables. But how is Secchi depth related to lake morphometry?

A regression model presented for Secchi depth gives mean annual Secchi depth (Sec_{12} in m) as a function of mean annual TP concentrations (TP_{12} in µg/l), mean annual lake color (Col_{12} in mg Pt/l) and mean depth (D_{mv} in m); $r^2 = 0.82$ (see Håkanson and Boulion, 2002). This model is graphically illustrated in fig. 4.28 and given by:

$$\log(Sec_{12}) = 0.86 - 0.0456 \cdot \sqrt{Col_{12}} - 0.114 \cdot \log(D_{mv}) - 0.143 \cdot \log(TP_{12}) \quad (4.11)$$

From fig. 4.28, it is evident that lake morphometry (D_{mv}) influences Secchi depth very much. This influence is also shown in table 4.18 using the data from appendix 8.3. From this table, one can note:

- Mean depth (D_{mv}); a very good predictor ($r^2 = 0.46$), which describes the fact that internal lake processes (resuspension, internal loading, etc.) affect Secchi depth very much: The larger the mean depth, the larger the Secchi depth.

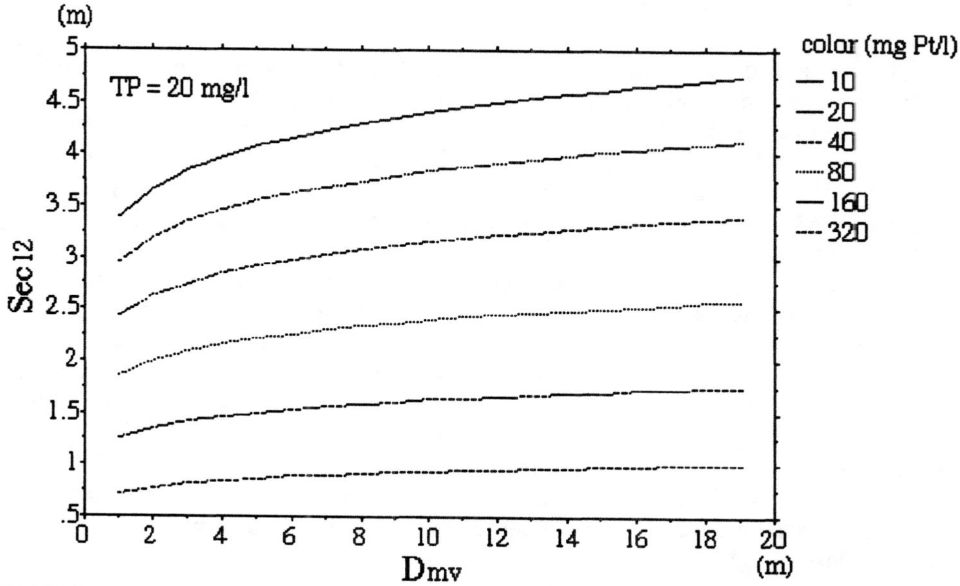

Fig. 4.28. Nomogram illustrating how Secchi depth (mean annual values in m) is related to the mean depth (D_{mv} in m) and lake color (mg Pt/l). In this plot, the lake TP concentration is set to 20 µg/l.

- The next most important lake morphometric factor is the theoretical lake water retention time (T). By accounting for T, r^2 increases to 0.51. This result is logical since Secchi depth is highly related to SPM (see fig. 4.23), which by definition settles out in lakes. So, lake morphometry is important for Secchi depth, not just in relation to resuspension, but also to sedimentation, which precedes resuspension. There are many compensatory and seemingly conflicting arguments to explain the role of morphometry in predicting Secchi depth. Secchi depth depends on color, which has a low sedimentation rate; Secchi depth depends on TP, because TP increases primary production of living and dead particles; Secchi depth depends on SPM, which settles out in lakes and reduce Secchi depth; Secchi depth depends on resuspension, which increases Secchi depth and depends on sedimentation. The ranking based on statistical analyses of empirical data is objective, and the results cannot be falsified by more data, which can only clarify the domain of the empirical model.
- What predictive power could be obtained by using just catchment parameters? From table 4.18B, one can note the r^2 value after three steps is as high as 0.46, which is lower than the result after two steps using lake morphometric parameters.
- Rock% enters at step 1, a logical positive relationship. If Rock% increases, the catchments have more infertile areas from which the transport of colored substances, suspended inorganic materials and nutrients would be small. It should, however, be noted that many parameters are related (causally, operationally and/or functionally) to each other and could replace one another in models of this kind without significant loss in predictive power. It is often difficult to give clear-cut causal explanation as to why and how a parameter like Rock% enters at this particular step in the regression.
- The ratio A/ADA is also a positive correlation. This result is logical: large lakes with small catchments generally have clear waters.

- Water discharge (Q_{mv}) has a positive r value, which is more difficult to explain.
- When morphometric and catchment parameters are pooled, the r^2 value after four steps is 0.65 (table 4.18C).

Numerous factors in a lake and its catchment can potentially influence Secchi depth. It is easy to speculate and qualitatively discuss such relationships. With these data, it is possible to quantitatively rank such influences and derive predictive models based on just a few, but the most important factors influencing among-lake variability in mean Secchi depth. Secchi depth depends on both allogenic and autogenic factors. The model based on map parameters could (statistically) account for 65% of the variability. This percentage is high in relation to the highest r_r^2 of about 0.99 (see table 4.2).

A "natural" Secchi depth could be estimated directly from the model based on map parameters. This prediction would account for differences in catchment area and lake morphometry. If the actual mean long-term Secchi depth differs from this predicted natural value, such divergences may be discussed in a quantitative manner. These empirical models only apply for lakes of the same type and within the domains of the variables used in the analyses. In other words, empirical models like these are not generally valid. But within their range of application, empirical models often give better predictions than dynamic models (Ahlgren et al., 1988; Peters, 1991). Furthermore, it is likely that the same basic principles apply for most lakes, although the parameter constants, and also the model parameters, may have to be altered to predict Secchi depth for lakes of other types.

Table 4.18. Ladder for the mean Secchi depth of lake water based on data from 36 months [$y = \log(Sec_{36})$; Sec_{36} in m]. The table lists the r^2 values and the models at each step. F > 4. Data from 86 lakes

Step	Parameter	r^2	Model
A. Sec_{36} vs. morphometric parameters			
1	$\log(D_{mv})$	0.46	$y=0.154+0.424 \cdot x_1$
2	$T^{0.1}$	0.51	$y=-0.209+0.314 \cdot x_1+0.445 \cdot x_2$
B. Sec_{36} vs. catchment parameters			
1	$\log(1+Rock\%)$	0.15	$y=0.301+0.124 \cdot x_1$
2	$(A/ADA)^{0.3}$	0.39	$y=-0.020+0.165 \cdot x_1+0.695 \cdot x_2$
3	$Q_{mv}^{0.1}$	0.46	$y=-0.563+0.171 \cdot x_1+0.892 \cdot x_2+0.578 \cdot x_3$
C. Sec_{36} vs. morphometric & catchment parameters			
1	$\log(D_{mv})$	0.46	$y=0.154+0.424 \cdot x_1$
2	$\log(OA)$	0.55	$y=0.344+0.439 \cdot x_1-0.172 \cdot x_2$
3	$(A/ADA)^{0.3}$	0.59	$y=0.198+0.418 \cdot x_1-0.137 \cdot x_2+0.279 \cdot x_3$
4	$\log(1+Rock\%)$	0.65	$y=0.023+0.354 \cdot x_1-0.075 \cdot x_2+0.451 \cdot x_3+0.097 \cdot x_4$

4.4.2.6. Sedimentation in lakes

The following factors are known to be important for sedimentation and, hence, also for the age of the active sediments in lakes, which regulates the fundamental transport process of burial, the transport of matter from the lake biosphere to the geosphere (Hilton, 1985; Evans and Håkanson, 1992):

1. A load factor describing the natural load of allochthonous materials and the anthropogenic load (from industries, urban areas and agricultural activities) on a given lake.
2. A pre-trapping factor describing the fact that lakes "downstream" receive less allochthonous materials than lakes "upstream", which act as sediment traps for suspended particles/aggregates.
3. A production factor describing the autochthonous particle production (the internal bioproduction).
4. A deposition factor describing the capacity of a given lake to act as a sediment trap (the larger the lake volume and the longer the water retention time, the higher the entrapment capacity if all other factors are held constant).
5. A resuspension factor linked to the morphometry of the lake, the topographical characteristics of the lake surroundings and the wind climate.

Evans and Håkanson (1992) demonstrated that sedimentation in lakes can be predicted from a model based on (1) mean pH as a water chemical indicator (of very many chemical, biological and physical processes in the lake and its catchment area), (2) maximum depth as a link to lake morphometry and the potential for resuspension, and (3) lake area as a measure of lake size and lake productivity.

This section gives an empirical model for gross sedimentation in lakes (see table 4.19). It is based on the same empirical data set used by Evans and Håkanson (1992). This model uses data on gross sedimentation from sediment traps placed in the deepest parts of the lakes (Sed_{DW}). Thus, the equation provides maximum values of gross sedimentation. Sedimentation increases from about zero at the limit between areas of transport and areas of accumulation to maximum values at the deepest part of the basin (sediment focusing; see fig. 4.20, left). In this approach, there is a correction factor applied to the model-predicted values based on this knowledge. The predicted values for gross sedimentation are more correct for a U-shaped basin, with a volume development ($V_d = 3 \cdot D_{mv}/D_{max}$) of 3, and too low for a V-shaped basin with a V_d smaller than 3. Thus, the ratio $V_d/3$ is used as a correction factor.

The model for gross sedimentation in lakes is:

$$GS = (V_d/3) \cdot Sed_{DW} \quad (4.12)$$

and Sed_{DW} is the gross sedimentation in sediment traps, as determined from the empirical model in table 4.19:

$$Sed_{DW} = 10^{(-1.56-0.53 \cdot \log(Drel)+0.4445 \cdot pH+0.8524 \cdot \log(TP))} \quad (4.13)$$

Table 4.19. An empirical model for gross sedimentation in near-bottom sediment traps (GS in µg dw/cm²·day) based on data from 25 lakes (from Håkanson, 2000; F > 4). D_{rel} = relative depth (= $(D_{max}\cdot\sqrt{\pi})/(20\cdot\sqrt{A})$; TP = mean lake TP concentration in µg/l.

Step	r^2	Model variable	Model
1	0.45	D_{rel}	$\log(GS)=2.05-0.70\cdot\log(D_{rel})$
2	0.66	pH	$\log(GS)=-0.75-0.68\cdot\log(D_{rel})+0.35\cdot pH$
3	0.79	TP	$\log(GS)=-1.56-0.53\cdot\log(D_{rel})+0.445\cdot pH+0.852\cdot\log(TP)$

This empirical model gave an r^2 value of 0.79 for these 25 lakes. The model applies if the model variables are within the following ranges:

	D_{rel}	pH	TP (µg/l)
Minimum	1.06	4.8	6.6
Maximum	1.86	6.6	35

The values for pH and TP should be the mean, characteristic lake values, for example, the mean annual values. Thus, gross sedimentation increases with:

1. Relative depth (D_{rel}); D_{rel} is directly related to internal loading (resuspension) and the form of the lake. The shallower the lake, the smaller the D_{rel} value, the larger the resuspension, and the larger the gross sedimentation. How the relative depth influences sedimentation is graphically illustrated in fig. 4.29.
2. Lake pH, which influences the aggregation of suspended particles, and many other lake processes (the higher lake pH, the more productive the lake and the higher the gross sedimentation).
3. Lake TP concentration, which is directly related to lake production.

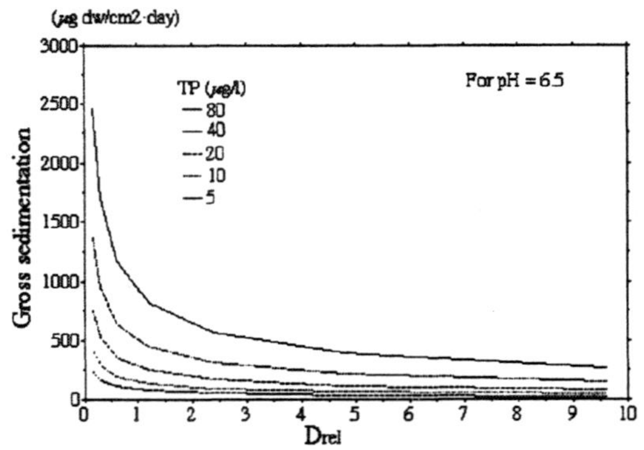

Fig. 4.29. Nomogram illustrating how gross sedimentation (mean annual values in µg dw/cm²·day) is related to the relative depth (D_{rel}) and lake TP concentrations (mg/l). In this plot, pH is set to 6.5.

If gross sedimentation (GS) is larger than 2000 µg dw/cm²·day, one can assumed that the oxygen consumption is so high that the oxygen concentration is smaller than 2 mg/l and bioturbation is halted because the meio- and macrobenthos die. So, sedimentation in lakes influences fundamental lake properties, such as burial and the survival of zoobenthos. Basically, oxygen consumption in lake sediments depends on the organic content of the sediments; and the next section will focus on this important variable.

4.4.2.7. Sediment organic content

Organic materials play many fundamental roles in aquatic ecosystems, as a carbon/energy source, as carrier particles for pollutants, as one of the most important regulating factors for the benthic habitat, etc. (Hansen, 1961; Wershaw et al., 1969; Mann, 1982; Wetzel and Likens, 1990). Lake sediments comprise (at least) three fundamentally different types of matter:

1. Autochthonous organic matter (like plankton) is produced in the lake. The average composition of planktonic materials is $C_{106}N_{16}P$, which gives a C/N ratio of about 5.6, and loss on ignition (IG; a standard measure of sediment organic content) is usually less than 20% dw (Håkanson and Jansson, 1983).
2. Allochthonous organic matter (like humus), generally (Golterman, 1975; Gjessing, 1976; Thurman, 1985) has a C/N ratio of 10-20 and a loss on ignition (IG) above 20% dw.
3. Minerogenic (inorganic) matter (like sand and silt), generally has a C/N ratio of 15 to 25 (Bowen, 1966) and a loss on ignition (IG) of less than 20% dw.

Fig. 4.30 also uses the classic classification of lakes in terms of autotrophy (on the x-axis) and allotrophy (on the y-axis) to highlight the fact that loss on ignition is about twice the carbon content (i.e., $IG \approx 2 \cdot C$).

The basic aim here is to look at the role of lake morphometry in predicting and understanding sediment organic content. Which are the most important, and the less important, parameters (x) for the variability among lakes in sediment IG, the target y-variable?

If the mean rate of sedimentation is 0.5 to 2 mm/year, the mean age of the uppermost sediment layer (0-1 cm) is 5-20 years. So, the mean IG determined from several areally distributed 1 cm thick sediment samples represents a mean value not just for the entire lake but also for a long period of time. This fact means that one cannot expect to successfully predict IG from water chemical variables, like pH, alkalinity, color and TP, which show great seasonal variability in a given lake.

Table 4.20 gives an empirical model for sediment IG based on the data discussed in appendix 8.3. The results of the stepwise multiple regression show:

- The best morphometric predictor of IG is the theoretical lake water retention time (T), which gives $r^2 = 0.2$.
- A much better regression model can be obtained for four catchment parameters, Till%, Mire%, catchment area (ADA), and relief (RDA). The r^2 value is as high as 0.77 (table 4.20B).
- If one morphometric parameter, the volume development (V_d) is added, the r^2 value increases to 0.83 (table 4.20C), which is very high. The highest reference r^2 (r_r^2) is 0.99 (from table 4.2).

So, sediment organic content (IG) is mainly linked to catchment parameters, but also, to a lesser extent, to lake morphometry.

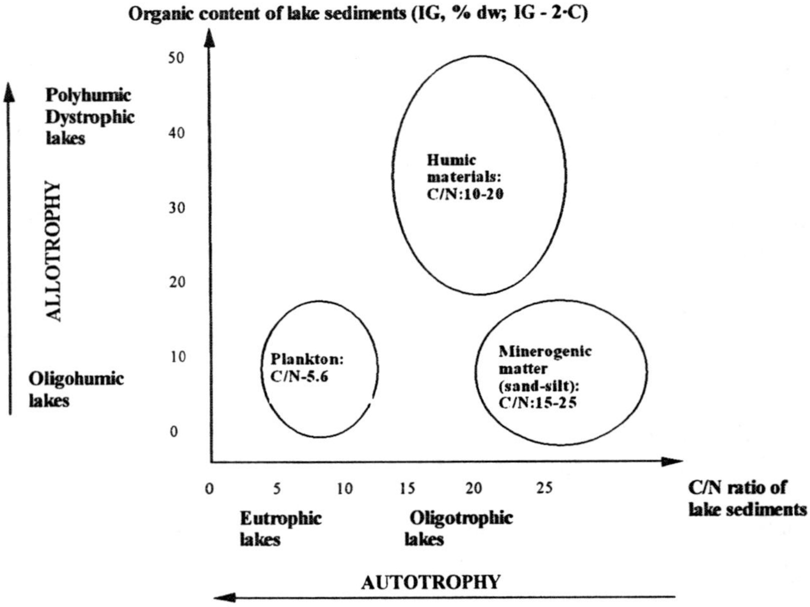

Fig. 4.30. Lake classification from the relationship between the C/N ratio and the loss on ignition (IG of surficial sediments (modified from Håkanson, 1995a).

Table 4.20. An empirical model to predict lake characteristic values of the organic content (= loss on ignition, IG in % dw) of surficial A sediments based on data from 33 lakes (data from Håkanson and Peters, 1995; F > 4).

Step	Parameter	r^2	Model
A. log(IG) vs. morphometric parameters			
1	log(T)	0.20	$y=36.6+10.7 \cdot x_1$
B. log(IG) vs. catchment parameters			
1	Till%	0.38	$y=9.07+0.357 \cdot x_1$
2	Mire%	0.55	$y=1.30+0.354 \cdot x_1+0.579 \cdot x_2$
3	log(ADA)	0.66	$y=11.14+0.279 \cdot x_1+0.604 \cdot x_2-6.879 \cdot x_3$
4	log(RDA)	0.77	$y=43.62+0.184 \cdot x_1+0.595 \cdot x_2-9.551 \cdot x_3-14.50 \cdot x_4$
C. log(IG) vs. morphometric & catchment parameters			
1	Till%	0.38	$y=9.07+0.357 \cdot x_1$
2	Mire%	0.55	$y=1.30+0.354 \cdot x_1+0.579 \cdot x_2$
3	log(ADA)	0.66	$y=11.14+0.279 \cdot x_1+0.604 \cdot x_2-6.879 \cdot x_3$
4	log(RDA)	0.77	$y=43.62+0.184 \cdot x_1+0.595 \cdot x_2-9.551 \cdot x_3-14.50 \cdot x_4$
5	V_d	0.83	$y=53.01+0.187 \cdot x_1+0.531 \cdot x_2-9.459 \cdot x_3-14.17 \cdot x_4-7.77 \cdot x_5$

4.4.2.8. Oxygen in deep water

An effect-load-sensitivity (ELS) model for coastal eutrophication is shown in fig. 4.31. The mean summer oxygen saturation in the deep water (the water beneath the wave base) is used as a operational effect variable (O_2Sat in %). The rationale for using this chemical measure as an operational effect variable is related to the linkages between the conditions for the bottom fauna, the oxygen conditions in the sediment/water zone and the load of organic matter to the sediments. When the O_2 saturation is lower than about 20%, key functional benthic groups are extinct. It is practical to have ELS models based on such a simple chemical variable as O_2Sat, but then one must demonstrate its biological/ecological significance, which is quite clear for O_2Sat (see also Cornett and Rigler, 1979). It should be noted that fig. 4.31 is included here to stress the role of morphometry in these contexts and that the data for this model emanate from coastal areas in the Baltic, which is a brackish water system.

The following working hypotheses were tested and verified when the model in fig. 4.31 was constructed (see Håkanson, 1999):

1. No single factor could explain the variability in mean O_2Sat that exists among these coastal areas. Several factors must be included and each factor is likely to add only a limited predictive power.
2. O_2Sat depends on both load factors (related to nitrogen and phosphorus concentrations in water) and sensitivity factors linked to the size and form (the morphometry) of the coast. In other words, the way the coast looks regulates how the coast functions.

It should be noted that the mean O_2Sat value is not a constant, but a variable, and that a model based only on morphometric parameters cannot be used for site-specific predictions of time-dependent y-variables. The empirical model in fig. 4.31 is based on nutrient concentrations and morphometry, and the aim is to predict long-term summer averages of O_2Sat for entire and well-defined coastal areas defined according to the method discussed in chapter 3. In fact, a very important consideration concerns the definition of the coastal area (where to place the boundaries toward the open water areas and/or adjacent coastal areas). It is crucial to use a technique that provides an ecologically meaningful and practically useful definition of the coastal ecosystem. The coastal boundaries are here operationally defined by means of the "topographical bottle-neck technique" to minimize the exposure (see fig. 3.58).

The results shown in fig. 4.31 give O_2Sat versus a load function plus morphometric parameters. One can note that O_2Sat may be predicted quite well from the load function ($TN+10 \cdot TP$) plus four morphometric (sensitivity) parameters: The higher the load of both N and P, the lower O_2Sat; the longer the theoretical deep water retention time (T_{DW}), the more sheltered the coast is (given by the filter factor, FF; see fig. 3.58 for definition), the deeper and larger the coastal area and the lower the filter factor, the lower O_2Sat. The r^2 value after five steps is 0.93.

Note again that in models of this kind one can often replace a model variable (x) with related variables from the same cluster. For example, FF may be replaced by the section area; the load function ($TN+10 \cdot TP$) could be replaced by other types of load functions. Since "everything is related to everything else" in ecosystems like these, it is generally difficult to give clear-cut causal explanations why a certain x-variable is linked to a certain y-variable. Regression models should be built upon simple, easily accessible, logical variables that show a minimum of interdependence.

The model in fig. 4.31 should not be used for, for instance, coastal areas dominated by heavy tides.

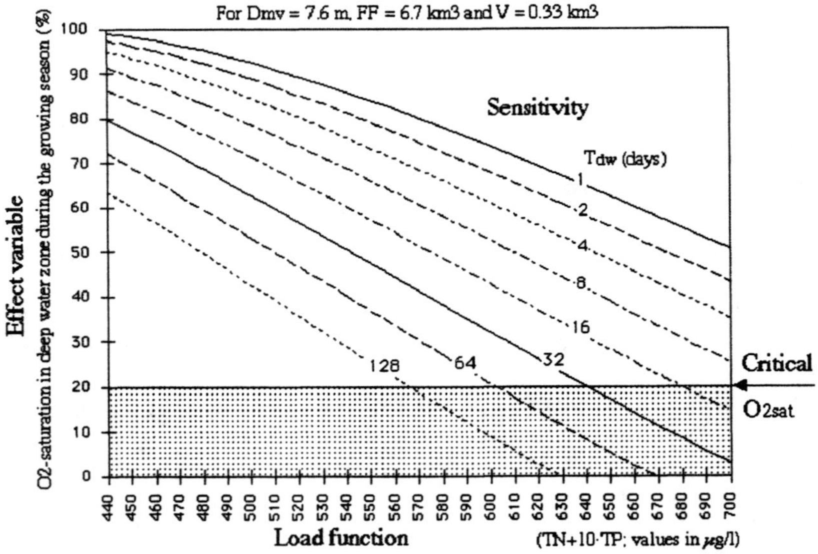

	r2-value
Step 1: Nutrient response function (TN+10·TP)	0.49
Step 2: Theor. deep water retention time (TDW)	0.76
Step 3: Mean depth (Dmv)	0.89
Step 4: Filter factor (FF)	0.91
Step 5: Coastal volume (V)	0.93

Data from 23 Baltic coastal areas

Model:

$$O2_{sat}=100\cdot SIN(14.83-4.46\cdot \log(TN+10\cdot TP)-0.40\cdot \log(1+T_{dw})-1.05\cdot \log(D_{mv})-0.021\cdot FF+0.275\cdot \log(V))$$

Fig. 4.31. Effect-load-sensitivity (ELS) model using the oxygen saturation in the deep water in Baltic coastal areas as the operational effect variable (modified from Håkanson, 1999).

When the zoobenthos die, the biological mixing (= bioturbation) of the sediments will halt (see Håkanson and Jansson, 1983). This cessation has profound consequences for the lamination of the sediments and hence also for all interpretations about the age of sediments and the age distribution. Generally, various types of zoobenthos live in sediments down to about 5-15 cm sediment depth (see fig. 4.32). The upper part of the sediment column is biologically active in the sense that the bottom fauna can influence the physical, chemical and biological conditions of the sediments and cause bioturbation. If the larger animals (macro- and meiofauna) die, bioturbation is halted and laminated (layered and unmixed) sediments may appear. The continuous sedimentation will cause the sediment layer to grow upward so that the bioturbation limit (the limit between the upper biological layer and the lower biopassive or geological layer) moves upward. Bioturbation will influence the age and the age distribution at any given sediment depth, with evident and profound implications in contexts of sediment dating, such as when the sediments are used as an historical archive. When bioturbation is negligible, for example, during anaerobic conditions, the age of the laminated sediments can be determined in a very straightforward manner by "counting annual varves."

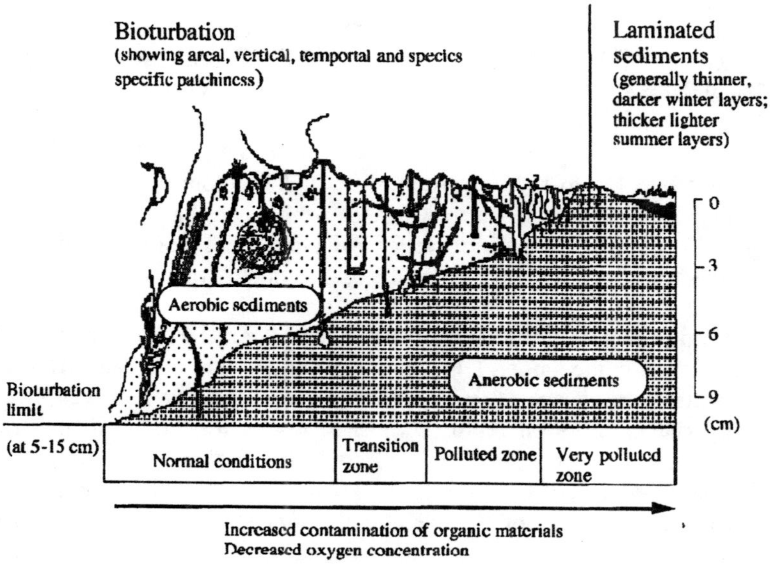

Fig. 4.32. Illustration of the relationship between deposition of organic matter, oxygen consumption, bioturbation and laminated sediments. Under aerobic (= oxic) conditions zoobenthos may create a biological mixing of sediments down to about 15 cm sediment depth (the bioturbation limit). If the deposition of organic materials increase and hence also the oxygen consumption from bacterial degradation of organic materials, the oxygen concentration may reach the critical limit of 2 mg/l, when zoobenthos die and bioturbation ceases and laminated sediments appear (figure modified from Pearson and Rosenberg, 1976).

4.4.3. Predicting toxins in biota

This section will discuss ELS models for toxic substances. There will be two examples, one for mercury, the other for radiocesium. There are similar examples in the scientific literature, but the aim here is not to make a literature review. It is to give two examples on the role of lake size and form in regulating effects of toxic substances.

4.4.3.1. Mercury in fish

Mercury is certainly one of the most studied environmental pollutants (see Lindqvist et al., 1991). There are about 60,000 publications dealing with mercury as an environmental pollutant. But contrary to common belief, Hg does not appear to be a major threat to aquatic life (but there are effects for, for example, birds). This difference in effect partly depends on the affinity of Hg to become bound to different types of carrier particles (clay minerals, humus and algae), whereupon the biological uptake of the metals is more difficult. Hg is, however, certainly a threat to the foetuses of pregnant women if the mother eats Hg contaminated fish. The operational effect variable, which has been used for Hg for many years, is the Hg concentration in 1-kg pike (*Esox lucius*). Pike is a stationary predator eaten by man. Obviously, the Hg concentration is not an effect variable in the same way as reproductive disturbance in roach or mortality in crayfish due to acidification (fig. 4.21), but it is the target operational effect variable for mercury in lakes.

The long-term environmental goal set by the several national environmental protection agencies is that the level of Hg in 1-kg pike (Hg_{pi}), and in all types of fish consumed by man, must not exceed 0.5 mg/kg ww. This means that most people would be able to eat fish from lakes without needing to consider health risks.

The empirical regression model used here comes from Håkanson (1999). The ladder is shown in table 4.21 and the ELS model in fig. 4.33. To develop this model data on lake area, mean depth, mean annual pH (pH_{12}), lake characteristic TP concentration TP_{12}), median Hg concentration of surficial sediments (0-1 cm), HgS, empirical Hg concentration in 1-kg pike (Hg_{pi}) were collected from 39 lakes. These lakes cover a fairly wide range in size (from 0.12 to 12.7 km^2), pH12 (from 4.8 to 6.4), TP concentrations (from 6.9 to 49.0 µg/l) and sediment contaminations of Hg (from 156 to 1300 µg Hg/kg ww). Lakes belonging to this type are likely to have fish with high Hg concentrations. Eutrophic lakes are likely to have fish with lower levels of mercury. The ranges in fig. 4.33 give the domain for the empirical model, and this model should not be used for lakes outside this domain, and not for other types of aquatic systems, like reservoirs.

From table 4.21, one can note that a lake morphometric parameter, the dynamic ratio (DR), is ranked in the pool position in the ladder. DR is more important than the load variable (HgS), and the water chemical sensitivity variables (pH_{12}) and TP_{12}. At the fifth step one can find yet another lake morphometric variable, the mean depth (D_{mv}). The reason for this position is related to the principles discussed in the interpretational key, mainly to the fact that mercury appears in lakes with a high particulate fraction (see Lindqvist, 1991). The carrier particles settle out in the lake and the internal fluxes are regulated by the specific morphometric characteristics of the lake, as this is expressed by the dynamic ratio and the mean depth. Fig. 4.33 demonstrates the important role that the mean depth plays in regulating the variations among lakes in this operational effect variable (Hg_{pi}). It is also interesting to note that the Hgpi values are higher in lakes with a low pH (see Håkanson, 1999, for further information on the mechanistic reasons for this), and in oligotrophic lakes (the "cleaner the water the more toxic the fish"). This conclusion may seem like a paradox, but it is a mathematical fact called "biological dilution", which is demonstrated here by empirical data. The mean concentration, C, of any toxic substance in lake fish (or any organism) is defined by the total amount of the toxin in fish (M) divided by the fish biomass (BM; C = M/BM). If there is an increase in fish biomass and no change in M, the concentration C must go down, and vice versa. The higher the lake TP concentration, the higher the primary, and hence also the secondary production, including the fish production.

Table 4.21. An empirical ELS model for the Hg concentration in 1-kg pike (Hg_{pi} in mg Hg/kg; from Håkanson, 1999), and the five steps in the stepwise multiple regression analysis. DR = the dynamic ratio; HgS = mean Hg-concentration in surficial sediments (0-1 cm; values in µg Hg/kg dw), pH_{12} = mean annual pH, TP_{12} = mean annual concentration of total phosphorus (µg/l) and D_{mv} = mean depth (m). Data from 39 lakes.

y-variable = $\log(Hg_{pi})$; F > 4

Step	r^2	Model variables	Model
1	0.63	$x_1=\log(DR)$	$y=-0.36-0.56\cdot x_1$
2	0.73	$x_2=HgS$	$y=-0.5-0.63\cdot x_1+0.000312\cdot x_2$
3	0.80	$x_3=pH_{12}$	$y=0.43-0.49\cdot x_1+0.000353\cdot x_2-0.14\cdot x_3$
4	0.82	$x_4=TP_{12}$	$y=0.5-0.41\cdot x_1+0.00351\cdot x_2-0.12\cdot x_3-0.00471\cdot x_4$
5	0.85	$x_5=\log(D_{mv})$	$y=0.4-0.44\cdot x_1+0.00397\cdot x_2-0.1\cdot x_3-0.00489\cdot x_4-0.15\cdot x_5$

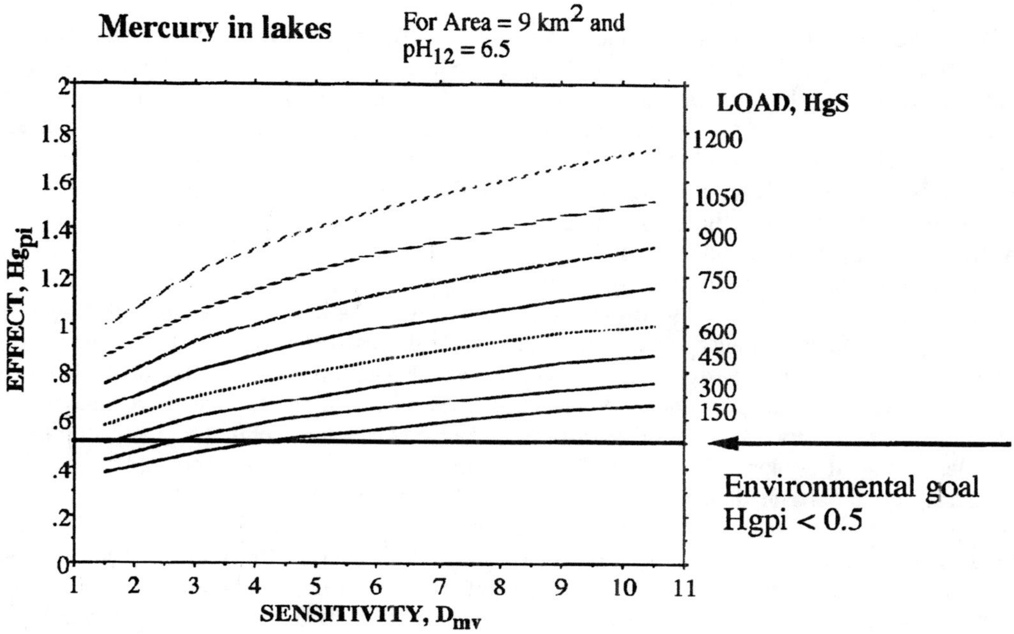

Model:

$\log(Hg_{pi})$=0.4 - 0.1·pH_{12} - 0.00489·TP_{12} - 0.44·log(DR) + 0.000397·HgS - 0.15·$\log(D_{mv})$

(r^2=0.85, n = 39)

Ladder:

Step 1. r^2=0.63. Sensitivity factor 1, dynamic ratio (DR)
Step 2. r^2=0.73. Load factor, HgS (µg/kg dw)
Step 3. r^2=0.80. Sensitivitiy factor 2, pH_{12}
Step 4. r^2=0.82. Sensitivity factor 3, TP_{12} (µg/l)
Step 5. r^2=0.85. Sensitivity factor 4, mean depth (D_{mv} in m).

Ranges of model variables:

$0.12 < A < 307$
$1.3 < D_{mv} < 10.6$
$4.8 < pH_{12} < 7.3$
$6.9 < TP_{12} < 49$
150<HgS<1300

Effect variable for Hg in lakes:

Hgpi= Hg-content in muscle in 1-kg pike; in mg Hg/kg ww

Fig. 4.33. ELS model for mercury in lake fish (1-kg pike) based on data from 39 lakes. HgS = median Hg content (µg Hg/kg dw) of surficial (0-1 cm) sediments; pH_{12} = mean annual lake pH; Total-P (= TP_{12}) = mean annual TP concentration (µg/l); D_{mv} = mean depth (m); DR = dynamic ratio; A = lake area (km^2).

The empirical model has been derived by means of stepwise multiple regression analysis. The r^2 value after five steps is 0.85 for $\log(Hg_{pi})$ and 0.83 for actual data on Hg_{pi}. The highest reference r^2 (r_r^2) for the actual values of Hg_{pi} is 0.96, since the characteristic CV for Hg in fish is about 0.25 (see table 4.2).

4.4.3.2. Radiocesium in fish

Connected to the Chernobyl nuclear accident in April/May 1986, large quantities of radiocesium (^{137}Cs) and radiostrontium (^{90}Sr) were released as a pulse. Following that signal through ecosystem pathways has improved the possibilities of identifying fundamental transport processes and structuring models (see Moberg, 1991; IAEA, 2000; Håkanson, 2000). The effects on aquatic organisms from ionizing radiation concern both individual and population levels, and the variables of interest are, for example, mortality rate, fertility rate and mutation rate (see Whicker and Schultz, 1982; Jimenez and Gallego, 1998). There are major differences between different organisms in sensitivity to ionizing radiation, and there are major differences for one and the same organisms living in different environments.

Environmental factors can also influence responses to radiation, for example, in mammals. Radioresistance increases under anoxic conditions. Salinity tends to increase radiosensitivity due to the metabolic stresses they demand on the organism. Radiation response also depends on the stage of the cycle the cell is in when the exposition occurs (IAEA, 1988). There are also indirect effects of radiation. For example, more resistant organisms can earn competitive advantage over others less resistant groups. The difficulties of getting meaningful results about radiation effects on natural populations in their natural environments have limited the available information on many species and communities, and very little information has been obtained about the interactive effects on radiation with other agents or stresses and the long-term effects of chronic low-level radiation. Aquatic organisms will receive external exposure from radionuclides in water and sediments and internal exposure from nuclides accumulated in their tissues. The calculation of such doses is the goal of several dosimetric models (IAEA, 1988), which use as input the amount of radionuclides in the water.

Lacking well-tested operational effect variables related to the reproduction or survival of key functional organisms in real aquatic ecosystems, the concentration of radiocesium in fish muscle consumed by man is used as an operational effect variable in the following ELS analysis. The guideline value for commercial marketing of fish recommended by governmental agencies, such as the Swedish National Food Administration, is 1500 Bq/kg ww.

An empirical ELS model ($r^2 = 0.92$) for radiocesium in lakes is given in table 4.22 (from stepwise multiple regression analysis) and illustrated in fig. 4.34. The guideline value is marked with an arrow. It should be noted that the model in fig. 4.34 only applies for a given year after the Chernobyl accident. The fallout onto the lake and its catchment area (Cs_{soil} in Bq/m^2) is the load variable and the operational effect variable is the Cs-concentration in pike caught in the spring of 1987, one year after the fallout (Cs_{pi87} in Bq/kg). The most important sensitivity factor in this model is lake conductivity. The more ions similar to Cs, like K, Ca, Na and Mg, the higher the conductivity and the lower the uptake of ^{137}Cs, a case of "chemical dilution" (Carlsson, 1978). Specifically, potassium (K) can be taken up in fish in almost the same way as radiocesium, and calcium can replace radiostrontium, so the more ions like K and Ca in the water, the less ^{137}Cs and ^{90}Sr will be taken up via cell membranes. This effect is "chemical" dilution. For a peak emission like this, the bio-uptake of ^{137}Cs, and the Cs concentration in fish, is lower in lakes with fast water turnover than in lakes with slow water turnover. This result represents a case of normal "water dilution", and it illustrates the importance of lake morphometry in these contexts. The opposite is valid for Hg, which is supplied to the lake continuously. In that case, the bio-uptake increases with increased runoff of Hg and water from the catchment, that is, when the Hg load from the runoff is renewed many times per time unit (Meili, 1991a, b). An increase in lake TP concentrations (in lake bioproduction) causes "biological dilution" for both Hg and ^{137}Cs (Håkanson and Peters, 1995). These relationships are quantified in the ELS model in table 4.22 and fig. 4.34. The latter gives theoretical water retention time on the x-axis to stress the role of morphometric parameters in this context.

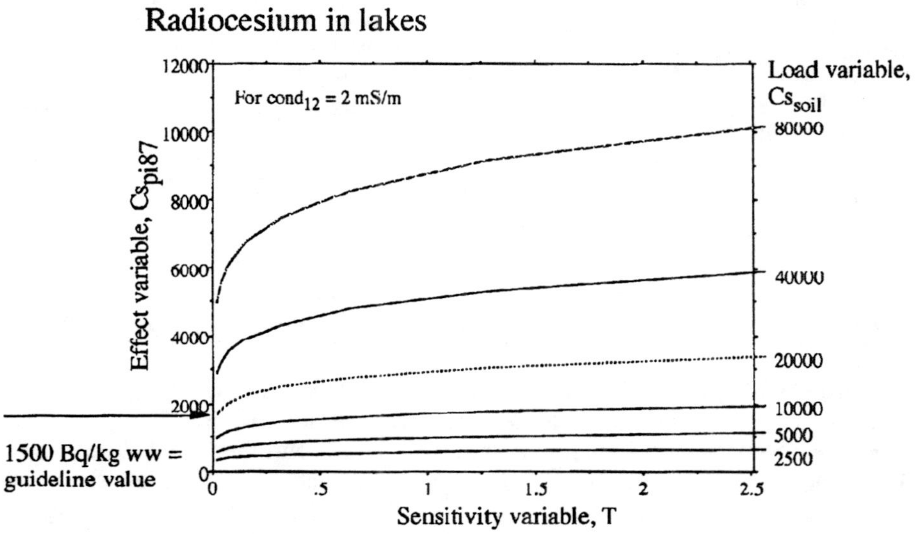

Model :
$\log(Cs_{pi87})=0.431+0.788\cdot\log(Cs_{soil})-1.155\cdot\log(cond_{12})+0.147\cdot\log(T)$

$(r^2=0.90;\ n=40)$

Effect variable: Cs_{pi87}; ^{137}Cs in pike caught in spring 1987 (Bq/kg ww)
Load variable: Cs_{soil}; fallout (Bq/m^2); range 2500-70,000
Sensitivity variables:
- $cond_{12}$; mean annual lake water conductivity (mS/m); range 1.6-8.2, and
- T; theoretical lake water retention time (yr); range 0.02-2.9

Fig. 4.34. ELS diagram for cesium in pike, which gives Cs_{pi87} (cesium in pike in spring 1987) as a function of Cs_{soil} (fallout of radiocesium after Chernobyl), mean annual lake water conductivity ($cond_{12}$) and the theoretical water retention time (T).

The highest reference r^2 (r_r^2) for radiocesium in fish is 0.97 (see table 4.2), and the r^2 value after five steps in table 4.22 is 0.92 [for $\log(Cs_{pi87})$].

Radiocesium seems to have a considerable particle affinity (Madruga and Cremers, 1997; Konitzer and Meili, 1997; Konoplev et al., 1997). This affinity is important and means that the behavior of radiocesium in lakes is strongly influenced by lake morphometry, as shown by the results in table 4.22. Also Hg has a strong affinity for particles (Lindqvist et al., 1991), which explains why the dynamic ratio is the most important factor regulating the variations in Hg concentrations in fish among lakes (see table 4.21).

Table 4.22. An empirical ELS model for the concentration of ^{137}Cs in pike (in this case fish caught in 1987 one year after the Chernobyl fallout; Cs_{pi87} in Bq/kg; based on data from Håkanson, 1991), and the five steps in the stepwise multiple regression analysis. Cs_{soil} = fallout (Bq/m^2), $cond_{12}$ = mean annual conductivity (mS/m), T = theoretical lake water retention time (yr), A_A = accumulation areas (% of A) and TP_{12} = mean annual concentration of total phosphorus (µg/l). Data from 40 lakes.

y-variable = $\log(Cs_{pi87})$; F > 4

Step	r^2	Model variables	Model
1	0.73	$x_1=\log(Cs_{soil})$	$y=-0.368+0.841 \cdot x_1$
2	0.87	$x_2=\log(cond_{12})$	$y=0.277+0.828 \cdot x_1-1.305 \cdot x_2$
3	0.90	$x_3=\log(T)$	$y=0.431+0.788 \cdot x_1-1.155 \cdot x_2+0.147 \cdot x_3$
4	0.91	$x_4=\log(A_A)$	$y=1.235+0.791 \cdot x_1-1.122 \cdot x_2+0.186 \cdot x_3-0.442 \cdot x_4$
5	0.92	$x_5=\log(TP_{12})$	$y=1.628+0.789 \cdot x_1-0.911 \cdot x_2+0.149 \cdot x_3-0.494 \cdot x_4-0.398 \cdot x_5$

Fig. 4.35 illustrates the role of lake morphometry in another way, not using the empirical model in table 4.22, which is static and ranks the factors which influenced the Cs concentrations in fish caught in lakes in 1987. The conditions then will never come back, but the factors and principles behind the regression may be interpreted in mechanistic terms and translated into a dynamic model (Håkanson, 2000), which may be used in simulating and forecasting. The simulations shown in fig. 4.35 use a dynamic model, and many of the features of that model will be discussed in chapter 5. Here, that model is used to illustrate what the Cs concentration in the water of Lake Zürich, Switzerland, could have been had this lake had a different volume, and hence a different theoretical lake water retention time ($T = V/Q_{mv}$). The default value for this lake is that T is 1 year, and fig. 4.35 shows the excellent fit between the empirical data and the modelled values when T is 1 year. What Cs concentrations would appear if the T value would have been 0.25, 0.5 or 10 years? This question is addressed in the simulations shown in fig. 4.35. One can note that if T had been shorter, the Cs concentrations would decrease faster, and vice versa. In radioecology and ecology, it is often interesting to predict "peak and tail" conditions related to a given single-pulse contamination; and fig. 4.35 gives one example of that.

It is obvious that many of the processes controlling the flow and biological uptake of Hg and ^{137}Cs (and other toxins) in lakes are controlled by morphometrical factors.

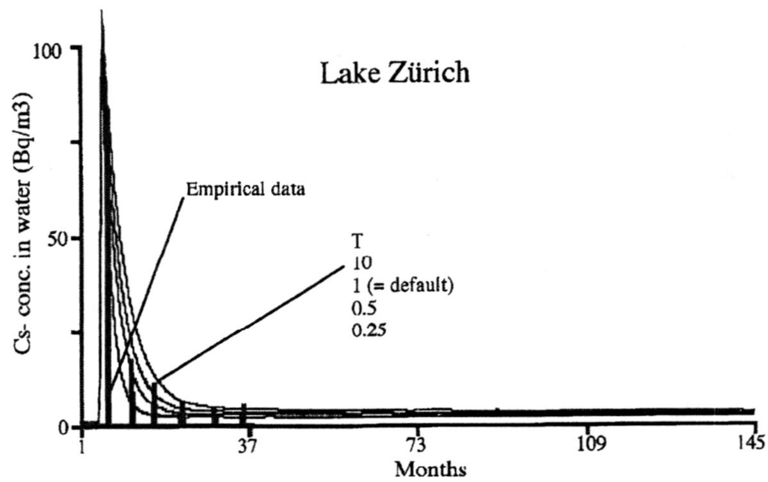

Fig. 4.35. A comparison between empirical data and modelled values for Cs concentrations in water in Lake Zürich (area = 537 km^2; mean depth = 90 m, maximum depth = 252 m) when the theoretical lake water retention time (T) has been changed from 0.25 to 10 years (1 is the default value). Model from Håkanson (2000).

4.4.4. Predicting key functional groups

This section will give two examples on the role of lake morphometry for two key functional groups, macrophytes and benthic algae (for more details related to the following two examples, see Håkanson and Boulion, 2002). The idea here has been to show that morphometry is important not only for water variables, such as Secchi depth and TP concentrations, but also for biological variables characterizing the lake foodweb. These functional groups are so called because changes in such groups will influence the structure and function of the entire lake ecosystem.

4.4.4.1. Macrophytes

Macrophytes can make a significant contribution to the total primary production, especially in shallow lakes. The macrophytes intercept nutrients and keep the nutrients bound for long periods. Consequently, they can influence water quality (Pokrovskaja et al., 1983). It is also important to emphasize that the evolution of any lake is closely connected with the overgrowing by rooted plants (Beeton and Edmondson, 1972). Macrophytes also provide an important protective environment for small fish. Although they may not be so important as a source of food for the fish, macrophytes can still influence fish production in lakes. The dominance of macrophytes or benthic algae depends on the characteristics of the bottom substratum. If it is stony, benthic algae dominate. If it is soft, macrophytes (+ epiphytes) are rooted and shade out the benthic algae. At higher latitudes, the littoral bottom areas are often covered by stony substratum and benthic algae. At lower latitudes, the littoral areas often consist of relatively soft sediments; and they tend to have a higher macrophyte cover.

Duarte and Kalff (1986) have quantified the relationship between littoral slope and macrophyte biomass for individual sites in lakes. The littoral slope ($slope_{lit}$ in %) in Lake Memphremagog, Canada, was measured as the angle between water and sediment surfaces using an echosounder; and the macrophyte biomass was given by the maximum values for the growing season in August. The equation giving maximum submerged macrophyte biomass (BM_{SMA}) from slope in this lake was:

$$BM_{SMA} = 124.47 + 990 \cdot slope_{lit}^{-0.81} \quad (4.14)$$
$$(r^2 = 0.73; n = 44)$$

Duarte and Kalff (1986) also tested this relationship for other lakes and obtained a high coefficient of determination ($r^2 = 0.81$).

It has been shown (see Scheffer, 1990; Scheffer et al., 2000) that in a given lake with macrophytes there can be two alternative stable states, one with turbid waters and few macrophytes, the other with clear water conditions and a high macrophyte cover.

The aim of this section is to discuss the role of morphometry for macrophyte production (PR_{MA}) and macrophyte cover (MA_{cov}) in lakes. For this purpose, data on macrophyte cover and production will be used from many lakes covering a wide domain of lake characteristics known or assumed to influence MA_{cov} and PR_{MA}. With these data, statistical analyses will be carried out to:

- Rank the factors influencing the variability among lakes in MA_{cov} and PR_{MA}. Note that within-lake variations will not be considered.
- Give statistical models to predict characteristic lake values of MA_{cov} and PR_{MA}.

Many of the lakes used here come from the former Soviet Union and have been presented by Håkanson and Boulion (2002). There are no reliable data on lake TP concentrations and pH from these lakes. This lack of data evidently sets limits to these models. It is interesting, though, to learn how far one can reach in terms of predicting macrophyte cover and production without such data. If phosphorus is a key limiting factor for MA_{cov} and PR_{MA}, then one should expect poor results.

The data base includes data on MA_{cov} and PR_{MA} from 35 large and small lakes (from 0.047 to 24000 km^2), from latitudes 13 °N (Lake Tchad in Africa) to 67.5 °N (Lake Big Kharbey, Vorkuta, Russia) and deep and shallow lakes (the maximum depth varies from 1.5 to 120 m). The macrophyte cover varies from 0.29% to 100%, and the macrophyte production from 1 to about 13,000 g ww/m^2·yr. The light conditions are important for the macrophytes; and the data base includes information on Secchi depth, which varies from 0.4 m to 6.7 m. In many lakes, the Secchi depth reaches its maximum value, the maximum depth.

Macrophytes have a characteristic growth pattern where the biomass attains maximum values in the summer and minimum values in the winter. Most shoots of the previous year are not kept at the time of the maximum biomass. Given this seasonal pattern, and the fact that non-respiratory losses are relatively small, the maximum seasonal biomass can be approximated as the cumulative annual net production. The dead macrophytes will be consumed by detrivores. Many studies have dealt with the definition of maximum macrophyte biomass (Westlake, 1980). Generally, those results apply to water bodies of the temperate zone, where the growing season for macrophytes lasts 4-5 months.

The areas above 1 and 3 m water depth (A_1 and A_3 in km^2) have been calculated using the approach discussed in chapter 3 because such areas are likely to have more macrophytes than deeper areas. To make these calculations, one needs data on lake area (A), maximum depth, D_{max}, and the volume development, V_d. A_1. The latter varies from 4% (0.04) to 53% among these 35 lakes.

As stressed, the slope conditions in the littoral zone have been demonstrated to be of great importance for the macrophyte biomass. Here, data on the mean lake slope given by eq. 4.15 were used:

$$\text{Tan(slope)} = D_{mv} /(0.165\cdot\sqrt{A}) \qquad (4.15)$$

One should also note that data on characteristic CV values for MA_{cov} and PR_{MA} are not available. Based on the existing information on characteristic CV values (see table 4.2), one can assume that a characteristic CV value for MA_{cov} should be rather high, about 0.3-0.45, and even higher for PR_{MA}, 0.35-0.5. This CV would mean that one cannot expect to get higher r^2 values in predictive models for MA_{cov} than 0.87-0.94 when modelled values are compared to independent empirical data, and no higher r^2 values for PR_{MA} than 0.84-0.92.

As stressed, most lake variables are lognormally distributed, implying that the logarithmic transformation would provide a frequency distribution close to the normal. The best transformation for latitude for these lakes is, however, ((90/(90-Lat)), yielding a MV/M_{50} ratio of 1.07 (see Håkanson and Boulion, 2002).

One can note from table 4.23 giving the ladders for MA_{cov} and PR_{MA} that the highest r^2 values are obtained for MA_{cov} rather than for PR_{MA}, which is in agreement with the assumed higher CV value for PR_{MA}. One can also note:

- Sec/D_{mv} is the most important factor to explain statistically the variation among these lakes in MA_{cov}; $r^2 = 0.52$.
- The next most important factor for MA_{cov} is latitude, which is obviously related to lake temperature. If latitude is added, 67% of the variation in MA_{cov} among the lakes can be accounted for statistically.
- The third factor is maximum depth; the deeper the lake, the smaller MA_{cov}, which is logical.
- The fourth factor is the area of the lake above 1 m, A_1; $r^2 = 0.84$.

- None of the other variables tested, including lake slope and ET areas, add significant predictive power to this regression model. The included model variables are not highly interrelated and the obtained r^2 value after four steps (0.84) is close to the assumed highest reference r^2 value for MA_{cov} of 0.87-0.94.
- Lake slope for whole lakes does not seem to be as important as littoral slope for individual sites, and this conclusion is understandable. About 75% of the variability within lakes in macrophyte biomass can be related to littoral slope, whereas lake slope only explains (statistically) less than 10% of the variation in MA_{cov} among these lakes and less than 2% of the variation in PR_{MA}.

Table 4.23. Stepwise multiple regression analyses using data from Håkanson and Boulion (2002) to calculate (A) the macrophyte cover (Mac_{cov} in % of lake area) and (B) the macrophyte production (PR_{MA} in g ww/m^2·yr). Sec = Secchi depth (m), D_{mv} = mean depth (m), Lat = latitude (°N), D_{max} = maximum depth (m) and A_{1a} = lake area above 1 m water depth (m^2).

A. y-variable = $\sqrt{Mac_{cov}}$; n = 19 lakes; F > 4

Step	r^2	x-variable	Model
1	0.52	Sec/D_{mv}	$y=1.944+4.825 \cdot x_1$
2	0.67	90/(90-Lat)	$y=6.757+3.83 \cdot x_1-1.57 \cdot x_2$
3	0.74	$\sqrt{D_{max}}$	$y=8.31+2.57 \cdot x_1-1.50 \cdot x_2-0.286 \cdot x_3$
4	0.84	$\log(A_{1a})$	$y=10.49+1.502 \cdot x_1-1.993 \cdot x_2-0.422 \cdot x_3+0.490 \cdot x_4$

B. y-variable = $\log(PR_{MA})$; n =35 lakes; F > 4

Step	r^2	x-variable	Model
1	0.73	$\log(Mac_{cov})$	$y=0.678+1.328 \cdot x_1$
2	0.80	90/(90-Lat)	$y=2.472+1.028 \cdot x_1-0.516 \cdot x_2$

Fig. 4.36 gives a 3D diagram relating the two most important model variables, Sec/D_{mv} and latitude, to MA_{cov}. The figure shows how these two x-variables influence MA_{cov} when D_{max} and A_1 are held constant. It is interesting to note the very strong influence of latitude on MA_{cov}.

The results of the statistical analyses for PR_{MA} are given in table 4.23B.

- 73% of the variability in PR_{MA} among these 35 lakes can be statistically explained by MA_{cov}. This result is logical.
- When the next most important factor, latitude, is added, the r^2 value increases to 0.80, which is close to the assumed highest reference value of 0.84-0.92.

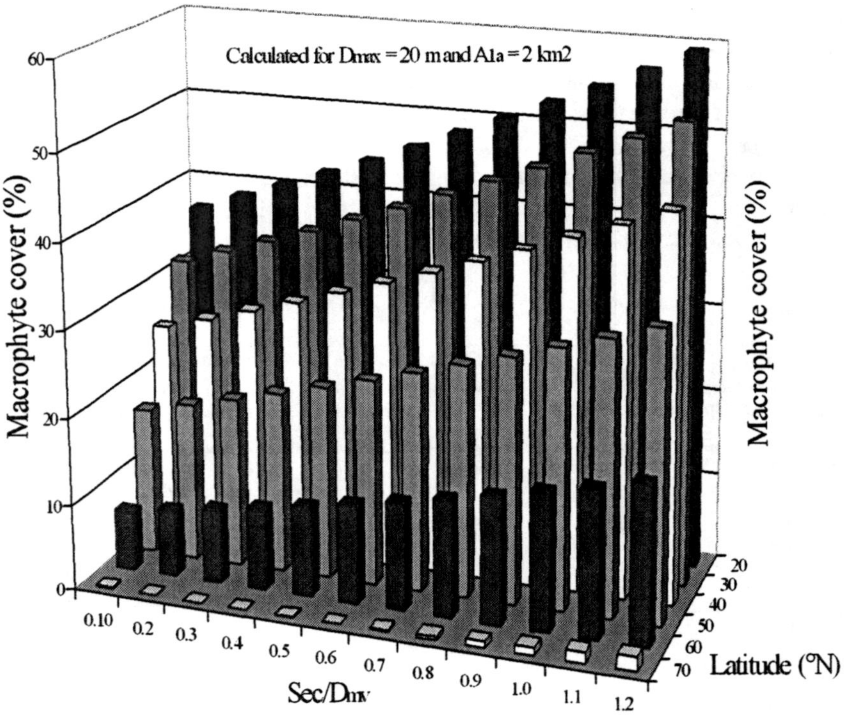

Fig. 4.36. 3D-diagram illustrating how the ratio Sec/D_{mv} and latitude influence macrophyte cover for a lake with a maximum depth of 20 m with the 2 km^2 lake area above 1 m water depth. From Håkanson and Boulion (2002).

4.4.4.2. Benthic algae

Zoobenthos, benthic algae and macrophytes are the three categories of key functional organisms related to sediments. Westlake (1980) also distinguishes between three groups of periphyton communities. One is attached to different substrates and forms dense belts including young, old and dead cells. The second group consists of numerous filaments or lumps of gelatinous materials. The third includes communities whose members are not strongly attached to the substratum. They are not aggregated and move freely over the substratum, generally the bottom sediments. Obviously, the horizontal and vertical distributions of benthic algae depend on the distribution of the illuminated substrates. In comparison with macrophytes and phytoplankton, benthic algae are confined to a relative thin surficial sediment layer, within which the concentration of cells can be very high. Generally, the term "periphyton" is applied for all forms of plants (with the exception of macrophytes) growing on submerged materials. The underwater substratum can be sediments, stones, boats, constructions and living organisms. Many epipelic (benthic) forms of algae display a daily migration pattern that creates additional variations in their distribution. The biomass of benthic algae is, however, generally expressed per unit of the water area, and not per unit of the substratum area (Loeb et al., 1983).

The littoral primary production often dominates the total lake primary production in shallow lakes. Deep lakes usually have only a relatively small part of the area within the littoral zone. A comparison of the littoral primary production in five oligotrophic lakes has indicated a large contribution of benthic algae (Loeb et al., 1983), but also a logical decreasing contribution of benthic algae to the total primary production with increasing water depth. Analyzing data for lakes from different regions in the former Soviet Union, Boulion (2001) has also shown that the contribution of benthic algae to the total littoral primary production depends on two important limiting factors, the morphometry of the littoral zone and the optical properties of the water.

Håkanson and Boulion (2002) have presented the data used here for benthic algae. There are data from nine lakes and 42 sampling sites. The information also includes, lake name, region, latitude, longitude, altitude, Secchi depth, mean lake slope, lake area, mean depth, maximum depth, sampling depth, phytoplankton production, production of benthic algae, "total" production (the sum of these two production values) and the requested target variable, the production of benthic algae (PR_{BA}) relative to the defined total production (PR_{tot}).

It is interesting to note that there is a very significant relationship (see fig. 4.37) between the ratio, PR_{BA}/PR_{tot}, for the individual sampling sites and the ratio between the sampling depth (D in m) and the Secchi depth (Sec in m):

$$(PR_{BA}/PR_{tot}) = -85.7\cdot(D/Sec)^{0.5} + 115.7 \quad (4.16)$$
$$(n = 42;\ r^2 = 0.75)$$

As much as 75% ($r^2 = 0.75$) of the variability among these 42 sites can be statistically explained by variations in the D/Sec ratio.

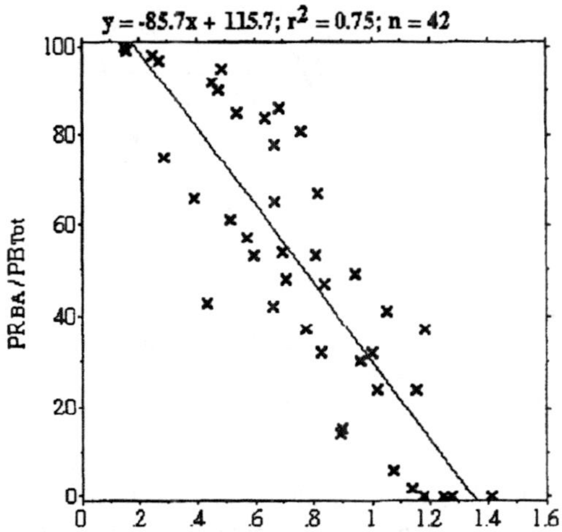

Fig. 4.37. The relationship between empirical data from individual sampling sites for, on the y-axis, the ratio between production of benthic algae to total production (of benthic algae plus phytoplankton), and on the x-axis, the ratio between sampling depth (D in m) and Secchi depth (Sec in m). From Håkanson and Boulion (2002).

4.5. Conclusion and compilation

This chapter has discussed regressions and ELS models and the role of lake morphometric parameters in such contexts. An interpretational key to relate variations within and among lakes in water chemical variables, operational effect variables, lake management variables and key functional groups of organisms have been discussed with a special focus on the role of lake morphometry. Fig. 4.38 is an attempt to summarize this information. It gives three categories in a lake-type triangle, with catchment parameters at the top, and morphometric parameters and "other causes" for variations at the base. All lake variables discussed here could be distributed in this triangle. This would tell us much about the factors causing variations in any given variable. Unfortunately, there is not complete information for many of the discussed variables. Conductivity, hardness (CaMg), calcium, iron, color, pH, alkalinity, Secchi depth and lake TP concentrations are, however, identified in the diagram.

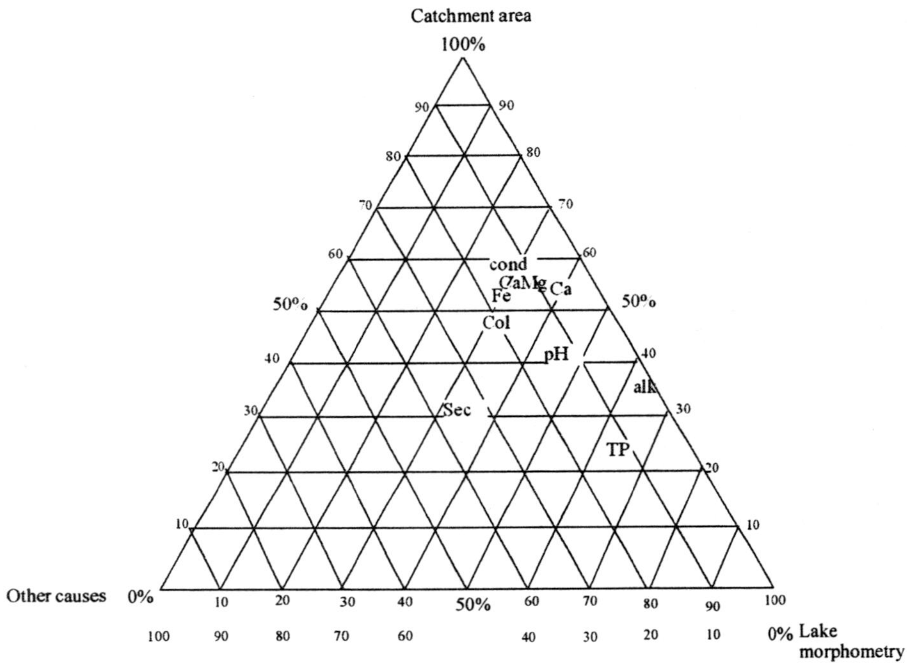

Fig. 4.38. Diagram showing how catchment area parameters (top), lake morphometric parameters (down, right) and other factors (down left; these factors include water chemical variables, climatological variables causing seasonal changes, etc.) influence predictions of the given water variables (conductivity, hardness = CaMg, calcium, iron, alkalinity, pH, total phosphorus and Secchi depth).

The calculations to obtain the data in fig. 4.38 are shown in table 4.24. To place a given lake variable in the triangle, one needs data on the maximum r^2 from regressions based only on lake morphometric parameters, maximum r^2 from catchment area parameters, maximum r^2 from both lake morphometric and catchment parameters and characteristic CV_w for the given variable so that the highest reference r^2 (r_r^2) can be obtained. The results show that 13% of the variability in mean annual values of lake conductivity may be linked to variations among lakes in morphometric parameters, 59% may be related to variations in catchment characteristics, 1% to inherence uncertainties in conductivity values and the rest, 27%, to other causes, mainly climatological factors.

Table 4.24. Compilation of data related to the highest reference r^2 (r_r^2) and the inherent uncertainties in the given variables [$=100\cdot(1-r_r^2)$], the maximum r^2 values obtained in the regressions based on lake morphometric parameters, catchment area parameters and pooled morphometric and catchment parameters and the corresponding percentages of the variations in the given lake variables [calculated as %morf = $100\cdot(r^2\text{morf}\cdot r^2\text{pooled}/(r^2\text{morf}+r^2\text{catch}))$]. "Other sources" means variations that may be related to climatological factors causing seasonal variations within lakes.

	r_r^2	r^2 from morf.	r^2 from catch.	r^2 pooled	% morf.	% catch.	% inherent uncertainty	% other sources
Conductivity	0.99	0.15	0.69	0.72	13	59	1	27
Hardness (CaMg)	0.99	0.13	0.67	0.71	12	59	1	28
Calcium	0.99	0.08	0.55	0.62	8	54	1	37
Color	0.97	0.28	0.69	0.68	20	48	3	29
pH	1	0.07	0.46	0.46	6	40	0	54
Alkalinity	0.92	0	0.34	0.34	0	34	8	58
Total-P	0.92	0.15	0.32	0.37	12	25	8	55
Secchi depth	0.99	0.51	0.46	0.65	34	31	1	34
Iron	0.96	0.1	0.69	0.71	9	62	4	25

From this diagram, one can note that among the discussed variables Secchi depth is the one most related to lake morphometry. The interpretational key explains why, and it has to do with the fact that Secchi depth in lakes is closely related to suspended particulate matter (SPM) concentrations, which by definition settle out in lakes.

5. The Importance Of Lake Morphometry In Limnology 2 – Mechanistic Roles In Dynamic Ecosystem Models

The aim of this chapter is to discuss and to clarify how parameters expressing elements of lake size and form influence lake processes. The first focus will be set on abiotic transport processes (sedimentation, resuspension, diffusion, burial, mixing and outflow) related to mass-balances. The second focus will be on how changes in abiotic factors influence key foodweb processes related to production, predation and biomasses of functional groups or organisms (phytoplankton, zooplankton, fish, etc.). To do this in a comprehensive way using empirical data from many lakes is far beyond the scope and possibilities of this book, or perhaps any publication. This chapter will do the second best, using mechanistically based validated models that have been tested with good results against empirical data from many lakes covering a wide limnological domain. The first part of this chapter will scrutinize the role of lake morphometry for the processes illustrated in fig. 1.1. Note, however, that this book is not meant as a textbook in modelling; rather, the idea is to try to explain the role of morphometric parameters in a simple, straight-forward manner without going into the details of the underlying differential equations. Maybe this is easier said than done?

5.1. Introduction

Predictive dynamic models are fundamental tools for forecasting the consequences of different remedial actions (see IAEA, 2000). The term "predictive model" is often used to indicate models in which one or a few important y-variables are predicted from a few x-variables that can be obtained easily from standard maps or monitoring programs. Many of the processes regulating the flow and biological uptake of substances in lakes are controlled by hydrological and morphometrical factors (Dahlgaard, 1994).

The following part concerns a rather comprehensive dynamic mass-balance model for substances (LakeMab). When validated (when the model output on concentrations in water, on suspended particles and in small planktivorous fish were tested against independent data from a wide domain of lakes), this model predicted very well. This predictive ability is exemplified in fig. 5.1 using data on radiocesium from the Chernobyl accident. When 357 empirical data from 23 very different European lakes were compared to modelled values of ^{137}Cs concentrations, the coefficient of determination (r^2) was 0.98 and the slope 1.04. This result is almost like an analytical solution.

Fig. 5.1 also gives the 95% confidence intervals for the mean y-value. One can note that the two confidence intervals are very close to the regression line. The 95% confidence interval for each individual data-pair is wider. For a normal distribution of random data, the standard deviation for the mean value (SD_{mv}) is $SD_{ind}/\sqrt{n}$, where SD_{ind} is the standard deviation for the individual data and n is the number of data used to calculate the mean.

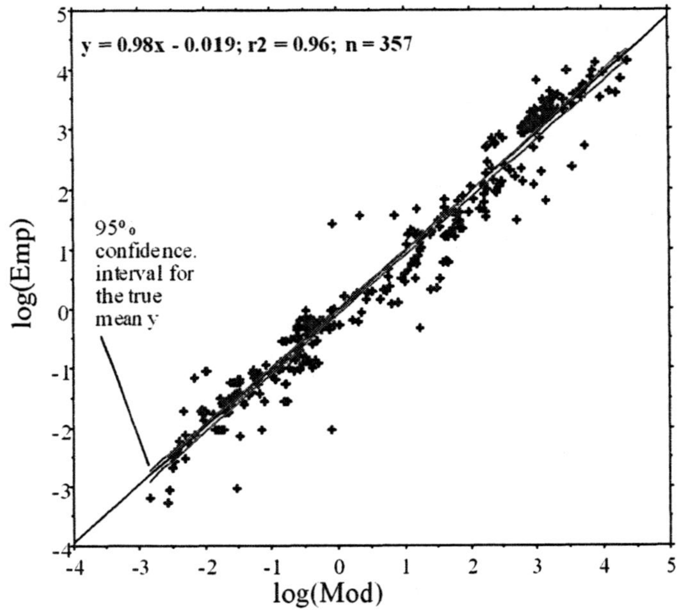

Fig. 5.1. Comparison between empirical data and modelled values for the mass-balance model for lakes (LakeMab) using radiocesium from the Chernobyl accident as a tracer (data from Håkanson et al., 2003). The figure also gives the regression line and the 95% confidence intervals for the true mean y. Note that the slope and the r^2 value are almost perfect, 0.98 and 0.96, respectively. The regression is based on 357 empirical Cs concentrations in water, on suspended particles and in small planktivorous fish from the 23 lakes covering a wide limnological range. Also note that there has been no tuning of the model. Only the obligatory lake-specific variables have been changed in this test.

Fig. 5.2 shows one of the many data series included in fig. 5.1. One can see that the modelled values in this validation are close to the mean empirical values and within the uncertainty bands of the empirical data. It can also be noted from fig. 5.2 that there are major uncertainties in the empirical data. The LakeMab model is not meant to predict the concentration in every sampling bottle; it is meant to predict the mean concentrations in surface water on a weekly or monthly basis. The time and area compatibility of the empirical data and the modelled values should be as high as possible; they should represent the same thing. This LakeMab model is based on a structure that is general and valid for all types of systems (rivers, coastal areas and lakes). The processes, such as inflow, sedimentation, resuspension, diffusion, burial and outflow, are general. There are also substance-specific components in some of the algorithms for the processes, which will be exemplified in the following section. This model has also been tested using sensitivity and uncertainty tests (Monte Carlo techniques) and validated (blind tested) with positive results for radiocesium from the Khystym accident (see Håkanson and Sazykina, 2001), for radiostrontium in lakes (see Håkanson et al., 2002), for calcium from lake liming (see Håkanson, 2003b; liming is a major environmental industry in many countries suffering from acid rain) and for phosphorus in lakes and hence also for lake eutrophication (Håkanson and Boulion, 2002).

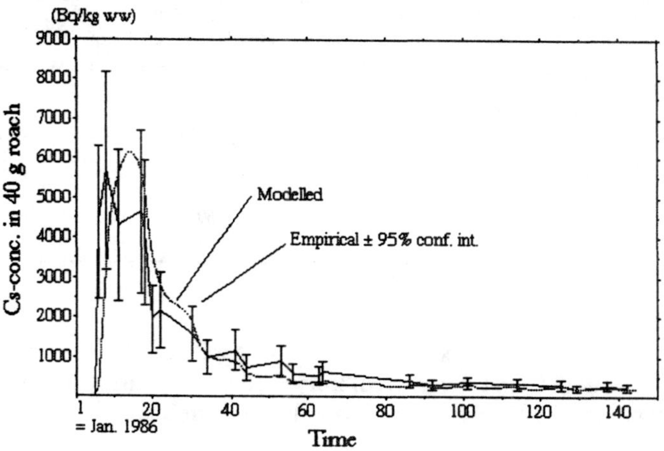

Fig. 5.2. The relationship between modelled values (from the LakeMab model) and empirical data for 40 g roach in Lake Siggefora, Sweden (data from Håkanson, 2000). The figure also gives the 95% confidence intervals for the empirical data (a measure of the uncertainty in the empirical data).

Modelling of phosphorus in lakes is a central paradigm in lake ecology and management; and fig. 5.3 gives an example of how the LakeMab model predicts TP concentrations in Lake Miastro, Belarus (see table 5.1 for data). It should be noted that to get the results in this figure there was no "tuning" of the model. There were no changes in any of the model algorithms of model constants, only changes in the obligatory lake-specific driving variables (given in table 5.1), which are readily accessed from standard monitoring programs and/or maps.

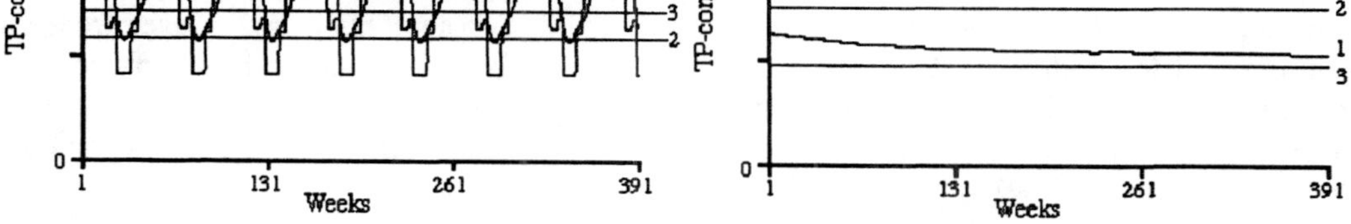

Fig. 5.3. Model validations for (A) TP concentrations in water and (B) TP concentrations in A sediments in Lake Miastro, Belarus. In fig. A there are three reference lines to the modelled values (given by curve 1), curve 2 gives TP concentrations from the OECD model (OECD, 1982), curve 3, TP concentrations from the Vollenweider model (Vollenweider, 1968) and curve 4, the empirical data for the lake. The empirical data are based on mean monthly values from the period 1980-90. Fig. B gives two empirical reference lines for the modelled TP concentrations in sediments, curve 2, the maximum TP concentration and curve 2, the mean value for this lake.

Table 5.1. Data from two lakes discussed in this chapter (miastro, Belarus, and S. Bullaren, Sweden). The data given here also the obligatory (lake-specific) driving variables needed to run the LakeMab and the LakeWeb models.

Lake	Catch-ment km^2	Area km^2	Mean Depth D_{mv}, m	Max. Depth D_{max}, m	pH	Color mg Pt/l	C_{in} μg, P/l	Lat. °N	Alt. m.a.s.l.	Precip . mm/yr
Miastro	133	13.1	5.4	11.3	8.0	31	80*	54.5	165	650
S. Bullaren	199	8.3	10.1	26.2	7.0	54	50	60.0	100	850

The fact that the LakeMab model has been validated with good results for many substances and for many types of lakes should lend some credibility to the following parts, which aim to highlight the role of lake morphometry in dynamic mass-balance modelling.

5.2. Mass-balance modeling

This modelling quantifies fluxes (mass/time), amounts (mass) and concentrations (mass/volume) to and from the defined compartments. Sedimentation will be treated first.

5.2.1. Sedimentation

To stress the role of lake morphometric parameters and factors highly dependent on lake morphometric parameters, all such x-variables will appear in boldface in the following presentation of the different equations (but not in the running text).

Obviously, inflow of water and/or substances to a lake cannot possibly be influenced by lake morphometric characteristics, so inflow will not be treated in the following text. Bio-uptake of substances and retention in biota are not treated because these processes do not depend directly on lake morphometry, only indirectly, since lake morphometry influences many water variables, such as pH, conductivity, hardness, Ca concentration, color and TP, which in turn are know to influence both biouptake and concentrations in biota (see chapter 4).

Basically, sedimentation is given by a mass (M in g) times a sedimentation rate (R_{sed} in "per time unit"; 1/time). The sedimentation rate can be influenced and modified by several factors, some of which are morphometrical. The aim here is to highlight such influences.

There are four compartments in the LakeMab model: Surface water, deep water, ET areas (areas where processes of erosion and transport dominate the bottom dynamic conditions) and accumulation areas (A areas). Also note the abbreviations: F for fluxes (g/month), C for concentrations (g/m^3), R for rates (1/month), M for masses (= amounts in g). Fluxes from one compartment to another are denoted, for example, F_{SWDW} for sedimentation from surface water (SW) to deep water (DW). The resuspended matter can be transported either back to the surface water (F_{ETSW}) or to the deep water (F_{ETDW}). How much will go in either direction is regulated by a distribution coefficient (DC) calculated from the form of the lake.

Sedimentation of suspended particles depends on:

1. A default settling velocity, v_{def}, which is substance-specific and related to the type of carrier particle. Examples are 12 m/yr for illite clays, the carrier particles for radiocesium (Håkanson, 2000) or 72 m/yr for planktonic materials (Håkanson and Boulion, 2002). The v value is changed into a rate (1/month) by division with the mean depth for the surface-water areas (D_{SW}) for sedimentation on the ET areas above the wave base and by

the mean depth of the deep-water areas (D_{DW}) for sedimentation in deep-water areas (the A areas) below the wave base.

2. The concentration of suspended particulate matter (SPM in mg/l) will influence the settling velocity. The greater the aggregation of suspended particles, the bigger the flocs and the faster the settling velocity (Kranck, 1973, 1979; Lick et al., 1992). This relationship is expressed by a dimensionless moderator (Y_{SPM}, see eq. 5.3).
3. The amount of resuspended matter will settle faster (two times faster in this model) than the primary materials. The resuspended particles have already been aggregated and they have also generally been influenced by benthic activities, which will create a "gluing effect". The fraction of resuspended matter (DC_{res}; the resuspension fluxes divided by all inflow fluxes) is calculated from eq. 5.4.

The mean depth of the surface water compartment is given by:

$$\mathbf{D_{SW}} = \mathbf{D_{wb}}/2 \tag{5.1}$$

The depth of the wave base, D_{wb} in m, is calculated according to eq. 3.19 and the mean depth of the deep-water compartment by:

$$\mathbf{D_{DW}} = (\mathbf{D_{max}} - \mathbf{D_{wb}})/2 \tag{5.2}$$

Where D_{max} is the maximum depth.

$$\mathbf{Y_{SPM}} = (1 + 0.75\cdot(\mathbf{SPM}/50 - 1)) \tag{5.3}$$

The reference value for this moderator is SPM = 50 (mg/l), which is a high value for SPM in lakes. The amplitude value is 0.75. If SPM changes with a factor of 2, for example, from 3 to 6 mg/l, the sedimentation rate increases with a factor of 0.34/0.295 = 1.15.

The fraction of resuspended matter (DC_{res}) is different in surface water and deep-water areas. In surface-water areas, it is:

$$DC_{resSW} = F_{ETSW}/(F_{ETSW} + F_{in} + F_{DWSWx}) \tag{5.4}$$

F_{ETSW} = Resuspension from ET areas to surface-water areas (g x/month).
F_{in} = Total inflows to the lake (from the catchment and direct fallout to the lake surface; g X/month).
F_{DWSWx} = Upward mixing, transport from deep water to surface-water areas (see later; g X/month).

For deep-water areas, DC_{res} is given by:

$$DC_{resDW} = F_{ETDW}/(F_{ETDW} + F_{SWDWx} + F_{ADW} + F_{SWDW} \tag{5.5}$$

F_{ETDW} = Resuspension from ET areas to deep-water areas (g X/month).
F_{SWDWx} = Downward mixing, transport from surface water to deep-water areas (see later; g X/month).
F_{ADW} = Diffusive transport from bioactive A sediments to deep-water areas (g X/month).
F_{SWDW} = Sedimentation from surface to deep-water areas (g X/month).

Sedimentation from surface water to ET areas is:

$$\mathbf{F_{SWET}} = M_{SW} \cdot \mathbf{R_{SW}} \cdot \mathbf{ET} \tag{5.6}$$

Where ET is the fraction of ET area ($ET = A_{ET}/A$). Sedimentation from surface water to deep water is:

$$\mathbf{F_{SWDW}} = M_{SW} \cdot \mathbf{R_{SW}} \cdot (1 - \mathbf{ET}) \tag{5.7}$$

and sedimentation from deep-water areas on A areas:

$$\mathbf{F_{DWA}} = M_{DW} \cdot \mathbf{R_{DW}} \tag{5.8}$$

The sedimentation rates for surface and deep-water areas are $R_{SW} = v/D_{SW}$ and $R_{DW} = v/D_{DW}$, respectively. The mean depths of the surface and deep-water areas, D_{SW} and D_{DW}, are calculated from eqs 5.1 and 5.2. The actual monthly settling velocity (v) is then calculated from the default settling velocity (v_{def} in m/yr) accordingly:

$$v = (v_{def}/12) \cdot \mathbf{Y_{SPM}} \cdot ((1 - \mathbf{DC_{res}}) + 2 \cdot \mathbf{DC_{res}}) \tag{5.9}$$

Where the resuspended fraction settles two times faster than the primary materials.

The boldface factors in the given equations demonstrate that size and form elements of lakes are of great importance for sedimentation in lakes.

5.2.2. Resuspension

The materials settling on the ET areas will not stay permanently where they were deposited; they will be resuspended by advective processes (mainly wind/wave activity). If the age of the material is T_{ET}, the resuspension rate may be set to ($1/T_{ET}$). In practice in ecosystem modelling, however, the retention rate is instead often set to $1.386/T_{ET}$, where 1.386 is the half-life constant and T_{ET} is the "full-time" age of the substance in the given compartment (here the ET areas). The half-life constant is basically from the equation for radioactive decay, $A(x) = A(0) \cdot e^{-kt}$; $A(t) = 0.5$, $A(0) = 1$ gives after logarithmation $\ln(0.5) = -k \cdot 0.5$ or $k = 0.694/0.5 = 1.386$.

If the age of the materials deposited on ET areas, T_{ET}, is set very long, (for example, ten years), these areas will function as A areas. If the age is set to one week or less, they will act as erosion areas. Generally for lakes, it is assumed that the mean age of these deposits is about one year (Håkanson and Peters, 1995). This value is also used as a default value in the LakeMab model. The approach used here to estimate the ET areas (and the presupposition for resuspension) is to calculate ET from the wave base (D_{wb}) and the form of the lake (V_d, see fig. 3.23). The ET value is used as a distribution coefficient. It regulates the sedimentation of matter either on A areas or on ET areas and hence also the amount of matter available for resuspension on ET areas. ET generally varies from 0.15, since there must always be a shallow shore zone where processes of erosion and transport dominate the bottom dynamic conditions (see fig. 3.25), to 1 in large and shallow lakes totally dominated by ET areas. In this modelling approach, however, ET is never permitted to become higher than 0.99, since one can assume that in most lakes there are deep holes, macrophyte beds or sheltered areas that would function as A areas. Also note that the wave base is never permitted to be lower than the evident boundary condition, the maximum depth (D_{max}). The boundary value for the wave base is that it must be $\leq 0.99 \cdot D_{max}$.

Resuspension back into surface water, F_{ETSW}, is then given by:

$$F_{ETSW} = (M_{ET} \cdot (1 - V_d/3) \cdot 1.386)/T_{ET} \quad (5.10)$$

Resuspension from ET-areas to deep-water areas, F_{ETDW}, by:

$$F_{ETDW} = (M_{ET} \cdot (V_d/3) \cdot 1.386)/T_{ET} \quad (5.11)$$

Where

M_{ET} = The total amount of resuspendable matter on ET areas (g).
V_d = The form factor; note that $V_d/3$ is used as a distribution coefficient to regulate how much of the resuspended material from ET areas will go either to the surface water or to the deep water.
T_{ET} = The age of the given substance in ET areas. As already explained, T_{ET} is set to one year, as a default value in the LakeMab model.

5.2.3. Stratification and mixing

The volume of the surface water (V_{SW}) is calculated by the following equation:

$$V_{SW} = V_{SW1} + V_{SW2} = V - A_{wb} \cdot (D_{max} - D_{wb})/3 \quad (5.12)$$

Where A_{wb} is the area below the wave base (the accumulation areas). A_{wb} (in m^2) is calculated from the equation defined from the hypsographic curve (see fig. 3.23) as:

$$A_{wb} = (A \cdot ((D_{max} - D_{wb})/(D_{max} + D_{wb} \cdot EXP(3 - V_d^{1.5})))^{(0.5/Vd)}) \quad (5.13)$$

To calculate A_{wb}, and hence also $A_{ET} = A - A_{wb}$, one needs data on the lake area (A), the maximum depth (D_{max}), the wave base (D_{wb}) and the form factor V_d,($V_d = 3 \cdot D_{mv}/D_{max}$, where D_{mv} = the mean depth).

The mixing between surface water and deep water depends on stratification, which in turn depends on many climatological factors (prevailing winds, season of the year, etc.). The following approach for mixing is simple. It gives the monthly mixing rate (R_{mix}; 1/month) as a function of the absolute difference between mean monthly surface and deep-water temperatures (SWT and DWT).

$$\text{If } ABS(SWT - DWT) < 4\ (°C), \text{ then } R_{mix} = 1; \text{ else } R_{mix} = 1/ABS(SWT - DWT) \quad (5.14)$$

The mixing rate is never permitted to be higher than 1. Mixing from deep water to surface water will cause a transport given by:

$$F_{DWSWx} = M_{DW} \cdot R_{mix} \quad (5.15)$$

Mixing from surface water to deep water is given by:

$$F_{SWDWx} = M_{SW} \cdot R_{mix} \quad (5.16)$$

It is well known that water (and air) temperatures are governed by many complicated climatological relationships. A method to estimate water temperature from map parameters on altitude, latitude and continentality is given in appendix 8.2.2. It has proven to give good predictions of mean monthly temperatures in lakes. Note that seasonal variability of both surface and bottom-water temperatures may be predicted by this approach. In many contexts, it may be possible to replace such modelled temperatures by measured values.

5.2.4. Diffusion

The second return flow from sediments back to water is diffusion. The diffusion rate for many substances (such as phosphorus and radiocesium) depends on sediment redox conditions (see Håkanson and Jansson, 1983). In oxic sediments, the diffusion rate is generally low. Note that the amount of a given substance X in the active A sediments may be high, so the diffusive flux of X can be relatively high even if the diffusion rate is low. If the sediments go anoxic, the diffusive flux for, for example, TP will increase significantly. The diffusive flux from A sediments to deep-water areas, F_{ADW}, is given by:

$$F_{ADW} = M_A \cdot \mathbf{R}_{ADW} \qquad (5.17)$$

Where the diffusion rate for phosphorus, R_{ADW} (per month), is given by:

$$\mathbf{R}_{ADW} = (0.0003/12) \cdot \mathbf{Y}_{Sed} \cdot \mathbf{Y}_T \cdot (DWT/4) \qquad (5.18)$$

Where 0.0003 per year is the default value of the diffusion rate for phosphorus. There are three dimensionless moderators in this equation:

1. Y_{Sed}, which expresses that lakes with high gross sedimentation are likely to have a higher oxygen consumption and lower HOC levels (hypolimnetic oxygen concentrations) than other lakes,
2. Y_T, which expresses that lakes with long theoretical water retention times ($T > 0.5$ yr) are likely to have lower oxygen renewal of the deep water, and hence lower hypolimnetic oxygen concentrations, and higher diffusion, than other lakes (if all else is constant), and
3. The ratio between predicted deep-water temperatures (DWT) and the standard reference temperature related to lake stratification (+ 4 °C), which expresses the fact that the bacterial degradation (= mineralization) of organic matter is temperature dependent.

The two dimensionless moderators Y_T and Y_{Sed} are given by:

$$\text{If } \mathbf{T} < 0.5 \text{ (yr), then } Y_T = 1; \text{ else } \mathbf{Y}_T = (1 + 1 \cdot (\mathbf{T}/0.5 - 1)) \qquad (5.19)$$

If $T = 0.5$ years, then $Y_T = 1$. An amplitude value of one means that if T is long, say 10 years, then $Y_T = 20$, and the diffusion rate is 20 times higher than in lakes with $T = 0.5$.

$$\text{If GS} < 50 \ (\mu g \text{ dw cm}^{-2} \cdot d^{-1}; \text{ see table 4.19), then } Y_{Sed} = 1;$$
$$\text{else } \mathbf{Y}_{Sed} = (1 + 1.35 \cdot (\mathbf{GS}/50 - 1)) \qquad (5.20)$$

Hence high hypolimnetic oxygen concentrations and low diffusion of phosphorus can be expected in lakes with gross sedimentation (GS) lower than 50 µg dw $cm^{-2} \cdot day^{-1}$. For lakes with higher GS, the diffusion rate will increase; and for GS = 2000 (which is very high and the limit where bioturbation is likely to halt), $Y_{Sed} = 54$, and the diffusive flux is of great importance in the mass-balance for TP in lakes.

For radiocesium, the diffusion rate (per month) is given in a similar but simpler way by:

$$\mathbf{R_{ADW}} = (0.002/12)\cdot(\mathbf{GS}/100)\cdot(\mathrm{DWT}/4) \quad (5.21)$$

As before, if GS > 2000 μg/cm^2·day, then the sediments are likely to go anoxic; and the default diffusion rate (0.002 per year) is simply multiplied by a factor of 20 and eq. 5.21 becomes:

$$\mathbf{R_{ADW}} = 20\cdot(0.002/12)\cdot(\mathbf{GS}/100)\cdot(\mathrm{DWT}/4) \quad (5.22)$$

It should be stressed that GS is a function of the relative depth (D_{rel}), lake pH and TP. Lake morphometry is very important for the GS value.

5.2.5. Burial

The age of the active A sediments (T_A) is needed to calculate the transport from active to passive A sediments, the burial flux, $F_{bur} = M_A \cdot R_{bur}$, where M_A is the mass of the substance in the compartment defined as active A sediments (M_A is automatically calculated in the LakeMab model) and R_{bur} is the retention rate for burial (dimension 1/time). The inverse of the age of the active A sediments may be used as a first approximation of the retention rate for burial. The age (T_A) is calculated from the ratio between the depth of the active sediments (D_{AS}, usually in cm) and the deposition (Sed_A, calculated from GS by dimensional adjustments) of materials on A areas (in cm/yr). Setting a default value for the depth of the active layer to 10 cm, and assuming that the sedimentation of matter can be calculated from GS in table 4.19, one can define the retention rate for burial (R_{bur}) by:

$$\mathbf{R_{bur}} = 1.386/\mathbf{T_A} \quad (5.23)$$

where

$$\mathbf{T_A} = \mathbf{BF}\cdot D_{AS}/\mathbf{Sed_A} \quad (5.24)$$

BF is the bioturbation factor (dimensionless), D_{AS} is the depth of the active A sediment layer (cm) and Sed_A is the mean sedimentation of matter on active A sediments (cm/month).

The biological mixing from the zoobenthos, gas ebullition, etc., will create also an upward transport of old, previously deposited sediments into the compartment of the surficial A sediments. This mixing will influence the age of the A sediments (T_A). Zoobenthos can eat the bottom sediments many times (up to seven times; see Håkanson and Jansson, 1983). When new materials are deposited on the sediments, they will be compacted and the water content will decrease with sediment depth. For this reason, the actual age of the active sediments is generally older than the value indicated by the ratio between the thickness of the active sediment layer (in cm) and sedimentation (in cm/yr). The correction factor for bioturbation (BF) is also related to gross sedimentation (GS). If the deposition is very high, the bottom-living animals are less likely to be able to create complete mixing.

As stressed, if gross sedimentation (GS) is larger than 2000 μg dw/cm^2·day, the oxygen consumption is generally so high that the oxygen concentration will become smaller than 2 mg/l and bioturbation halted because the meio- and macrozoobenthos will die. Gross sedimentation may either be measured, for example, by sediment traps or from analyzing sediment core data for radiocesium or lead-210 (see Håkanson and Jansson, 1983), or it may be estimated from the empirical model shown in table 4.19. The mean sedimentation (= deposition) of A areas, Sed_A in eq. 5.24, is then calculated from empirical data on gross sedimentation (GS) determined from sediment traps.

The bioturbation factor (BF) is given by:

$$\text{If BF} < 1\text{, then BF} = 1\text{; else } \mathbf{BF} = (\sqrt{D_{AS}} - 0.1\cdot(\mathbf{GS}/100 - 1)) \qquad (5.25)$$

Where $\sqrt{D_{AS}}$ is the bioturbation and compaction factor, which depends on the depth of the bioactive A layer (D_{AS}). If D_{AS} is 9 cm, then the value of this factor is 3; 0.1 is the amplitude value and 100 is the norm value in the dimensionless moderator for the influences of GS on the bioturbation factor (BF). If GS increases from the norm value of 100 to 1000, the BF value changes (if D_{AS} = 10 cm) from $\sqrt{10}$ = 3.16 to (3.16 - 0.9) = 2.3; and the mean age of the sediments changes in the same way.

5.2.6. Outflow

The outflow accounts for the flow out of the lake of water and substances dissolved and suspended in the water, as well as for the balance between the precipitation of water on the lake surface and the evaporation of water from the lake.

$$Q_{out} = Q_{in}\cdot Y_{SWT}\cdot Y_{prec} \qquad (5.26)$$

Where

Q_{out} = the water transport out of the lake (m^3/month);
Q_{in} = the water transport to the lake (see appendix 8.2.1).

The dimensionless moderator for surface-water temperatures (SWT in °C) and evaporation is defined in the following manner:

$$\text{If SWT} < 9\ °\text{C, then } Y_{SWT} = 1\text{; else } \mathbf{Y_{SWT}} = (1 - 0.4\cdot((\text{SWT}/9) - 1)) \qquad (5.27)$$

Where 9 is the norm temperature related to the duration of the growing season (Håkanson and Boulion, 2002) and 0.4 is the amplitude value, meaning that Y_{SWT} = 1 if the actual value of the mean monthly surface water temperature (SWT) is 9 °C; Y_{SWT} will approach zero when the actual mean monthly surface water temperature approaches 33 °C. Then the outflow of water (Q_{out}) also approaches zero.

The dimensionless moderator for precipitation (Y_{prec}) is defined by:

$$\text{If Prec} < 650\text{ mm/yr, then } Y_{prec} = (1 + 1.8\cdot(\text{Prec}/650 - 1));$$
$$\text{else } Y_{prec} = (1 + 0.5\cdot(\text{Prec}/650 - 1)) \qquad (5.28)$$

Where 650 is the norm for the mean annual precipitation (Prec); 1.8 is the amplitude value if Prec is smaller than 650; 0.5 is the amplitude value if Prec is larger than 650 mm/yr. Eq. 5.28 means that Y_{prec} = 1 if Prec = 650; Y_{prec} will approach zero when the actual mean annual precipitation approaches 300 mm/yr, and then the outflow of water (Q_{out}) also approaches zero; Y_{prec} will increase when the actual mean annual precipitation increases. The outflow of water (Q_{out}) is twice as large when Prec is 2000 mm/yr compared to when Prec = 650 mm/yr.

The outflow for any dissolved or suspended substance depends on lake morphometry because the depth, volume and concentration of any substance X in the surface water compartment are calculated from lake morphometric parameters (wave base, area and form).

5.2.7. Water level fluctuations and resuspension

Water level fluctuations are common not only in many reservoirs, but also in natural lakes, depending on changes in land-use practices, precipitation and evaporation. The situation in Lake Aral is a famous example of changes in water level, but there are many less well-known situations, like Lake Kinneret, Israel (see Håkanson et al., 2000). It is evident that there are close links between changes in water level and lake morphometry. If the water level goes down, the lake area will be smaller, the wave base lower and the sediments and substances continuously accumulated below the previous wave base exposed and influenced by wind-induced wave activities. Resuspension is likely to increase for some time after the lowering of the wave base. Likewise, the concentration of suspended particulate matter (SPM) is also likely to increase. As a result, the Secchi depth will also be reduced, influencing the primary production of phytoplankton and benthic algae. If there are changes in primary production, there are also likely alterations in secondary production of zooplankton and fish. Such relationships will be treated in the next section dealing with lake foodweb interactions and the LakeWeb model. In this section, the focus is set on water level fluctuations, lake morphometry and resuspension.

Fig. 5.4 is included here to give a background to resuspension in lakes. The SPM concentrations in the water depend on the distance to the bottom. Close to the bottom, one should expect higher SPM concentrations (Floderus and Håkanson, 1989; Nõges et al., 1999) from resuspension. As discussed in chapter 3, the water depth also determines whether a bottom is dominated by erosion, transport or accumulation processes for cohesive fine particles. Shallow and wind/wave exposed sites/areas are generally dominated by processes of fine particle erosion and transport. The relationships between winds, waves and SPM concentrations have been investigated in several studies (Floderus and Håkanson, 1989; Hellström, 1991; Kristensen et al., 1992). A high wind speed might imply an increased SPM concentration. Not only is the prevailing wind situation of importance, but also the frequency of resuspensions (fig. 5.4). If there are many resuspensions per month, it is likely that there is less material on the bottom to be resuspended, compared to a situation with few resuspension events. Also the wind direction should be of interest. If there is a large fetch, the wave base (the water depth down to which the wave orbitals can resuspend fine particles) can be deep.

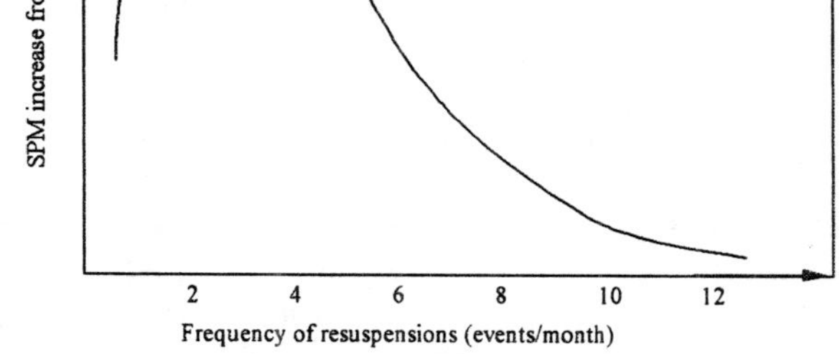

Fig. 5.4. Illustration of the relationship between the frequency of resuspension events and the increase in suspended particulate matter (SPM; note that there is no scale on the y-axis). Few resuspension events per time unit would imply a small increase in SPM in the water; many resuspensions would also mean a small increase in SPM because the amount of resuspendable matter would be small.

The changes in lake area and volume associated with changes in water level (D_{wl}) may be calculated from:

$$\text{If } D_{wl} = 0 \text{ (m), then } \mathbf{V} = \mathbf{D_{xmv} \cdot A_{max}};$$
$$\text{else } \mathbf{V} = (\mathbf{V_{max}}\ ((\mathbf{A_{max}}\text{-}\mathbf{A}) \cdot D_{WL}/2)\text{-}(\mathbf{A} \cdot \mathbf{D_{wl}})) \qquad (5.29)$$

Where V_{max}, A_{max} and D_{xmv} represent the volume, area and mean depth related to the highest water level as given by $D_{wl} = 0$. The mean depth (D_{mv} in m) associated with changes in D_{wl} from the highest level is given by the ratio V/A.

It should be noted that resuspension occurs from ET areas to surface water, from ET areas to deep water and from a lowering of the wave base both to the surface water and the deep water. Two distribution coefficients are needed to handle these fluxes. The resuspension from ET areas is distributed either to surface water or to deep water by means of the form factor, $V_d/3$, which is used as a distribution coefficient (as already mentioned); $V_d/3$ of the amount available for resuspension on the ET areas goes to deep water and $(1\text{-}V_d/3)$ goes to the surface water. When $D_{mv} \approx D_{max}$, the lake is U shaped and all resuspended matter from ET areas should go to the deep water (see chapter 3). If the distribution coefficient handling erosion related to a lowering of the wave base is set to 0.4, it means that 40% of the resuspended matter is assumed to be transported to the surface water and 60% to the deep water of the matter available when the wave base is lowered. To quantify this amount for a given lake is difficult, but the following approach may be used. The area exposed to erosion is calculated as the difference between the areas above the wave base at $D_{wl} = 0$ (A_{wb} in fig. 5.5) and above the wave base when the water level is lowered (A_{wbwl}). The larger the D_{wl} value, the larger the areas exposed to resuspension and the larger the erosion depth, which may be set to 5 cm as a default assumption. The erosion depth describes how sediments below the wave base will come above the wave base when the water level is lowered. These sediments may have been deposited a long time ago, and they are likely more compacted than recently deposited sediments. So, the lowering of the wave base may erode 5 cm of sediments, but this depth also depends on the change in the water level: A greater change in D_{wl} means a greater erosion depth. In short, the erosion depth (D_{er} in cm or m) should be a function of the changes in water level. In this model, it is simply given by:

$$\mathbf{D_{er}} = 0.05 \cdot \mathbf{D_{wl}} \qquad (5.30)$$

To calculate the amount of matter eroded, one can assume that the water content of the sediments exposed in connection to a lowering of the wave base is about 10% lower than the water content of the recently deposited matter (here set to W = 80% ww) and has a bulk density 10% higher than in recent sediments (set to d = 1.1 g ww/cm^3). Therefore, erosion (F_{ER} in tons dw/month) may be calculated as:

$$\mathbf{F_{ER}} = 10^{-6} \cdot (\mathbf{A_{wb}}\text{-}\mathbf{A_{wbwl}}) \cdot \mathbf{D_{er}} \cdot (10^6) \cdot d \cdot 1.1 \cdot (1\text{-}(W\text{-}10)/100) \qquad (5.31)$$

The erosion related to a lowering of the wave base can only continue as long as $D_{wl} > 0$. If the erosion rate (R_{er}) is set to 1/24, as a default assumption using monthly calculations, it means that all the material down to the erosion depth (D_{er}) will be eroded and a new equilibrium established after 24 months. If, for example, the erosion depth is 2 dm (related to a water level of 4 m beneath the maximum value), then it would take two years to erode (by wind-induced wave erosion) the material related to this lower wave base. If the water level is then raised, this type of erosion should stop.

Model for lake area vs. WL

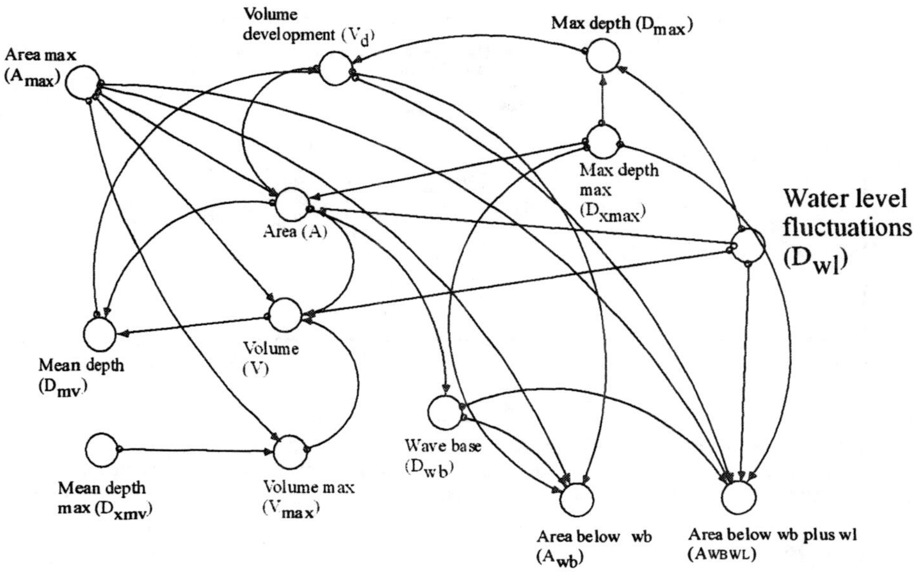

Equations:

$A_{max} = 170 \cdot 10^6\ m^2$

$D_{xmv} = 26\ m$

$V_{max} = 26 \cdot 170 \cdot 10^6\ m^2$

$D_{wb} = 45.7 \cdot \sqrt{(A \cdot 10^{-6})}/(\sqrt{(A \cdot 10^{-6})}+21.4)$

$D_{mv} = V/A$

$D_{max} = D_{xmax} - D_{wl}$

$V_d = 3 \cdot D_{mv}/D_{max}$

$A = A_{max} \cdot 0.01 \cdot (100 \cdot EXP(-3+V_d^{1.5}) \cdot (D_{xmax}-D_{wl})/(D_{wl}+D_{xmax} \cdot EXP(-3+V_d^{1.5})))\hat{}(0.5/V_d) \cdot 10\hat{}(2-1/V_d)$

$V = A \cdot D_{mv}$

$A_{wb} = A_{max} \cdot 0.01 \cdot (100 \cdot EXP(-3+V_d^{1.5}) \cdot (D_{xmax}-D_{wb})/(D_{wb}+D_{xmax} \cdot EXP(-3+V_d^{1.5})))\hat{}(0.5/V_d) \cdot 10\hat{}(2-1/V_d)$

$A_{wbwl} = A_{max} \cdot 0.01 \cdot (100 \cdot EXP(-3+V_d^{1.5}) \cdot (D_{xmax}-D_{wb}-D_{wl})/(D_{wb}+D_{wl}+D_{xmax} \cdot EXP(-3+V_d^{1.5})))\hat{}(0.5/V_d) \cdot 10\hat{}(2-1/V_d)$

Fig. 5.5. An outline of the model to calculate changes in lake area, volume, mean depth and maximum depth associated with water level fluctuations (data for Lake Kinneret, Israel; see also Håkanson et al., 2000).

Fig. 5.6 illustrates how this approach works. In this hypothetical example, the water level (D_{wl}) was suddenly lowered during month 12 in Lake Kinneret, Israel (ADA = 2560 km^2; A = 170 km^2; D_{xmv} = 26 m; D_{max} = 42 m; mean annual precipitation = 406 mm), to 4 m below the maximum value. The corresponding changes in lake area and SPM (in mg/l) can be seen by curves 2 and 3 and would lead to a marked increase in SPM associated with this change in D_{wl}. Such a change is likely to cause major effects on the structure and function of the lake ecosystem, and these effects will be treated in the next section.

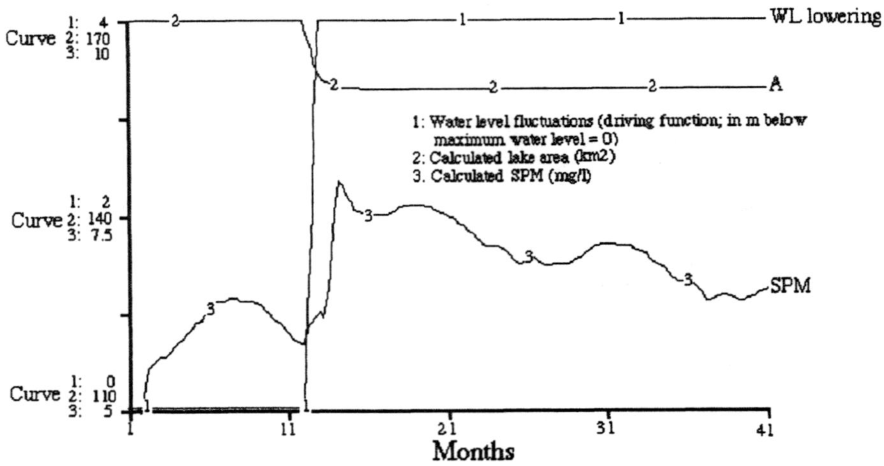

Fig. 5.6. Calculated consequences of water level fluctuations (a water level lowering as given by curve 1) in Lake Kinneret, Israel. Curve 2 gives the corresponding change in lake area and curve 3 gives the target variable for these model calculations, the changes in SPM values (from Håkanson et al., 2000).

5.3. Foodweb modelling

This section will use a lake foodweb model to exemplify how lake morphometry influences fundamental ecosystem functions. The LakeWeb model (Håkanson and Boulion, 2002) is a general model to quantify all important foodweb interactions and biotic/abiotic feedbacks, not for individual sites in lakes or for single species but for entire lakes and key functional groups. The model has been critically tested against comprehensive empirical data sets. The general set-up of the model is illustrated in fig. 5.7, and fig. 5.8 gives an example of the predictive power of the model along a trophic state gradient for herbivorous zooplankton (fig. 5.8A), predatory zooplankton (fig. 5.8B), prey fish (fig. 5.8C) and phytoplankton (fig. 5.8D). Similar tests have been carried out with the same positive results for all nine key functional groups of organisms included in the model - phytoplankton, bacterioplankton, benthic algae, macrophytes, zoobenthos, herbivorous and predatory zooplankton, prey fish and predatory fish. These are functional groups of organisms and not individual species, and they exist in most aquatic systems. Tests like those shown in fig. 5.8 have also been carried out with good results along pH gradients (acid status gradients), color gradients (humic level gradients), latitude and temperature gradients and lake size gradients. The LakeWeb model is based on many new approaches for the quantification of lake foodweb interactions, and some of those are given in appendix 8.5. Like the LakeMab model, it uses ordinary differential equations (the ecosystem scale) and gives seasonal (weekly or monthly) variations.

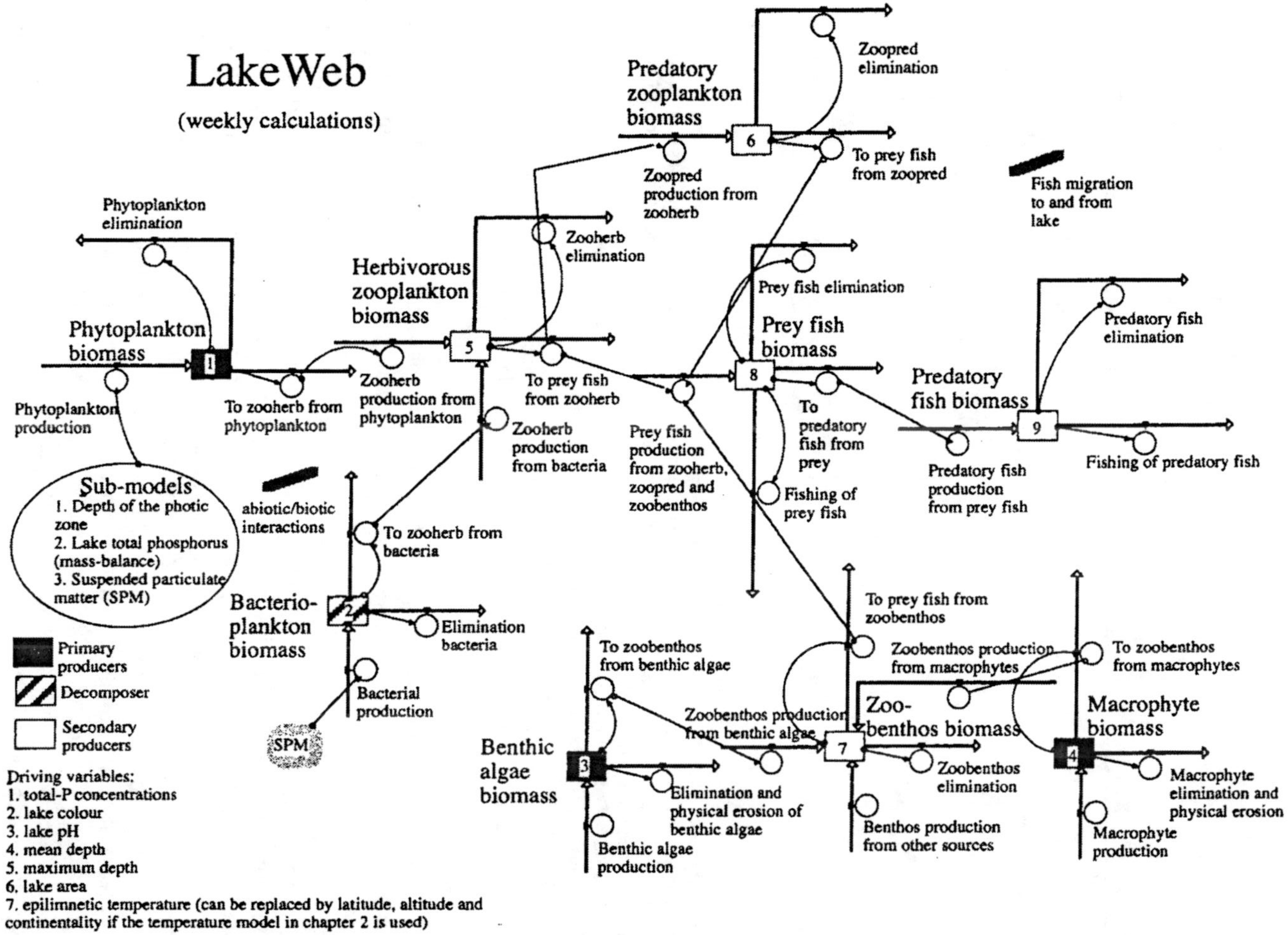

Fig. 5.7. Illustration of the LakeWeb model (from Håkanson and Boulion, 2002) with its nine groups of organisms (phytoplankton, bacterioplankton, benthic algae, macrophytes, herbivorous zooplankton, predatory zooplankton, zoobenthos, prey fish and predatory fish) and the seven obligatory driving variables. The model also includes sub-models for lake temperature that enables determinations of epilimnetic temperatures from latitude, altitude and continentality.

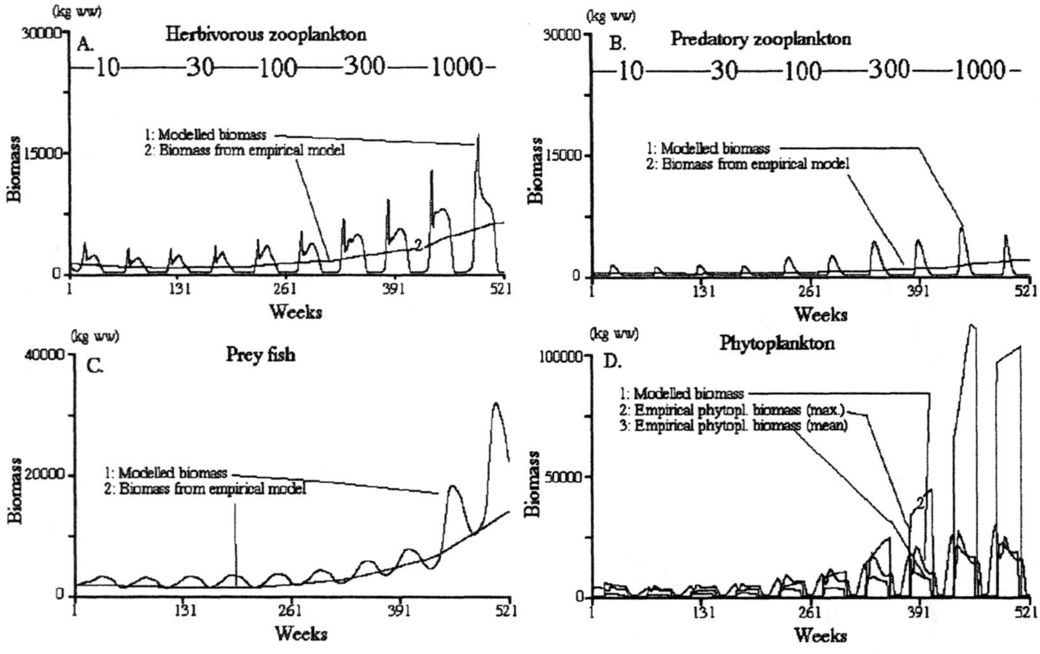

Fig. 5.8. Testing of the LakeWeb model for four variables under default conditions (lake area = 1 km^2, mean depth = 10 m, pH = 7) along a trophic state gradient. The driving variable is the tributary TP concentration, which has been varied in 2-year steps from 10 to 1000 μg/l, while all else is constant. Note that three of the empirical models (called norms) do not provide any seasonal patterns, only mean values for the growing season.

A: Gives predicted biomass of herbivorous zooplankton (curve 1) and the corresponding empirical norm (curve 2).

B: Gives modelled biomass of predatory zooplankton (curve 1) and the empirical norm (curve 2).

C: Gives biomass of prey fish (curve 1) and the corresponding norm (curve 2).

D: Gives predicted biomass of phytoplankton (curve 1) and the two empirical norms, the upper, maximum curve (2) as calculated for the entire lake volume, and curve 3 (the mean value) as calculated for the volume of the photic zone.

The LakeWeb model also includes a mass-balance model for phosphorus, the LakeMab model. By including the TP model in the foodweb model, it is possible to calculate the uptake and retention of phosphorus in biota.

In the following, the aim is to use the LakeWeb model, with its mass-balance model for phosphorus, to try to demonstrate the importance of lake morphometry for the structure and function of the lake ecosystem. The idea is to create a scenario where there is no change in the inflow of water and phosphorus to a system, and then test how different systems would react to the same phosphorus loading. The simulations will use data from Lake S. Bullaren, Sweden (see table 5.1). Then the mean depth of the lake will be changed in two steps (8.3 m is the actual value, and these simulations will also use a mean depth of 1.7 m and 41.5 m; the value is changed by a factor of 5), and the area of the lake will also be altered in two similar steps. Consequently, the response in a shallow and a deep, a large and a small lake will be simulated. It should be noted that changes in mean depth and area will influence other lake form parameters and many processes depending on these morphometric characteristics. These changes are accounted for in these simulations. The biotic target variables in the following simulations all depend on abiotic variables such as Secchi depth, SPM, TP concentration and lake pH. It is not possible to make appropriate interpretations of the results for the functional organisms unless the changes for these abiotic factors are first presented.

5.3.1. Changes in abiotic governing variables

Fig. 5.9 gives the results for lake TP concentrations in water (A) and in surficial A sediments (B). Fig. 5.9A first shows that for the initial period (the first 261 weeks, or 5 years), there is a good correspondence between the modelled values and the empirical mean (Emp. MV) TP concentrations for this lake. This figure also gives the corresponding data using the OECD model; and one can note that the OECD model predicts too low TP concentrations, which is understandable because there is also a significant internal loading of TP in this lake (see Håkanson, 1995b) and the OECD model does not quantify internal loading. Fig. 5.9B shows the good correspondence between modelled TP concentrations in A sediments and the empirical mean value. From a mass-balance perspective, it is important to note that the TP concentrations in water and sediments are close to the empirical data, which is an indication that the inflow, the internal process and the outflow are calculated correctly.

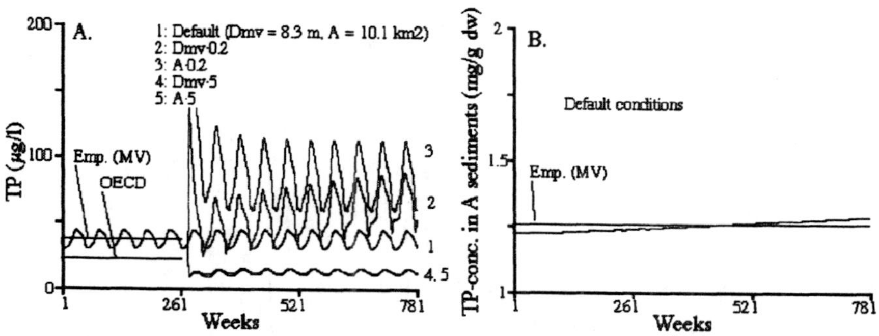

Fig. 5.9. Results for TP concentration in (A) water related to the changes in lake morphometry (mean depth set to 8.3 m, 0.2·8.3 = 1.7 m and 5·8.3 = 41.5 m; lake area set to 10.1 km^2 and to 0.2·10.1 = 2.02 and 5·10.1 = 50.5 km^2) and (B) TP concentrations in A sediments under default conditions compared to empirical sediment data for Lake S. Bullaren, Sweden.

In week 261, the changes occur suddenly, as if the nutrient loading entered five different lakes. One can see that for the small lake with an area of 2 km^2 (curve 3), the TP concentration would be very high, about 100 µg/l, which indicates hypertrophic conditions; for the shallow lake (mean depth = 1.7 m; curve 2), the TP concentrations would also be very high, about 75 µg/l. The actual lake (curve 1) has a TP concentration of about 35 µg/l, which is mesotrophic to eutrophic. Had the lake been deep or large (curves 4 and 5, respectively), the TP concentrations would have been about 10 µg/l, indicating oligotrophic conditions.

A lake's morphometry will have a very significant effect on how a given nutrient load will manifest itself in lake concentrations.

The corresponding results for Secchi depth (fig. 5.10A) and SPM fig. 5.10B) show:

- That there is a good correspondence between modelled Secchi depths and the value calculated by the empirical regression (in table 4.1), here called the norm. Also empirical Secchi depth data from the lake agree with these values.
- The deep lake (curve 4) will have a much higher Secchi depth, about 5 m, whereas the shallow lake (curve 2) would have a much lower Secchi depth, about 1.5 m. This result is related to resuspension, as well as to the TP concentrations and the corresponding linkages to primary production.
- As stressed (see fig. 4.26), there is a close causal inverse relationships between the Secchi depth and SPM; and fig. 5.10B demonstrates that the modelled SPM values are close to the values obtained by the empirical norms ("Norm Sec" from the regression

between Secchi depth and SPM and the "Norm" from table 4.1). The SPM values would have been very high indeed (about 30 mg/l) had this been a shallow lake (curve 2), and very low (curve 4) had it been deep.

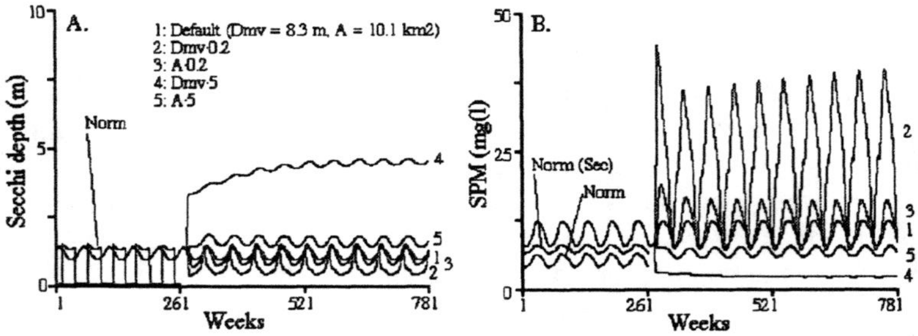

Fig. 5.10. Changes in Secchi depth (A) and concentration of suspended particulate matter (B) from the changes in lake morphometry (mean depth set to 8.3 m, 0.2·8.3 = 1.7 m and 5·8.3 = 41.5 m; lake area set to 10.1 km^2 and to 0.2·10.1 = 2.02 and 5·10.1 = 50.5 km^2; basic data from Lake S. Bullaren, Sweden).

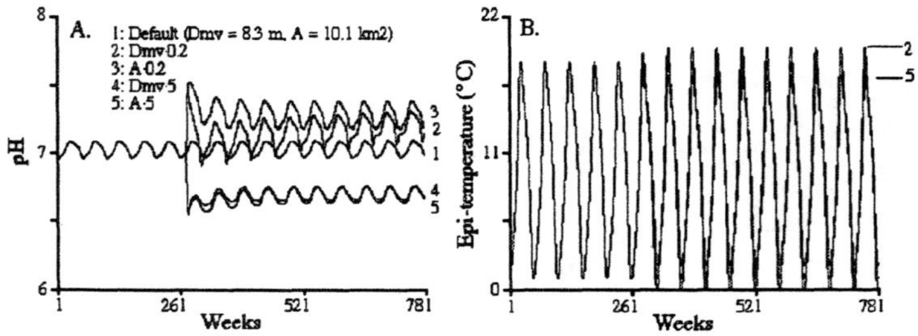

Fig. 5.11. Changes in mean weekly lake pH (A) and epilimnetic temperatures (B) from the changes in lake morphometry (mean depth set to 8.3 m, 0.2·8.3 = 1.7 m and 5·8.3 = 41.5 m; lake area set to 10.1 km^2 and to 0.2·10.1 = 2.02 and 5·10.1 = 50.5 km^2; basic data from Lake S. Bullaren, Sweden).

The changes in lake pH and surface water temperatures are given in fig. 5.11. One can see that (1) the initial pH is about 7, which is in good correspondence with empirical data (see Håkanson, 1995b). When TP concentrations and primary production are high (as in the shallow lake, curve 3, and the small lake, curve 2), pH becomes higher, and vice versa.

The changes in epilimnetic temperatures are not so large, but are important because lake temperature influences many foodweb processes, such as consumption and predation. Summer temperatures will be higher in shallower lakes (curve 2) and lower in large lakes (curves 4 and 5).

With the foregoing in mind, it will be interesting to see the corresponding results for biota.

5.3.2. Primary producers and bacterioplankton

The three primary producers in the LakeWeb model are phytoplankton, benthic algae and macrophytes. Bacterioplankton is a decomposer. The primary production of phytoplankton, benthic algae and macrophytes is directly related to abiotic limiting factors, such as the concentration of phosphorus, water clarity, temperature, pH and lake color, all of which depend on lake morphometry. This is accounted for in the LakeWeb model in various ways, generally by dimensionless moderator techniques. That is:

$$IPR_{PU} = PR_{PU} \cdot Y_X \cdot Y_{Sec} \cdot Y_{Temp} \quad (5.32)$$

Where

IPR_{PU} = the initial production of the primary unit; generally in kg ww per week;
PR_{PU} = the mass production rate; generally in kg ww per week;.
Y_X, Y_{Sec} or Y_{Temp} = dimensionless moderators expressing how changes in environmental conditions, like water chemistry (X), water clarity or Secchi depth, Y_{Sec}, and lake water temperature (temp) influence primary production. Y_X has the following general definition:

$$Y_X = (1 + amp \cdot (X_{act}/X_{norm} - 1)) \quad (5.33)$$

Where the amplitude value (amp) quantifies how changes in actual values (X_{act}) relative to a normal (= reference) value (= X_{norm}) influence the production; if $X_{act} = X_{norm}$, then $Y_X = 1$.

The change in biomass per time unit (dBM; in kg wet weight per week) of any primary producer (dBM_{PU}) is related to the initial production (IPR_{PU}), the loss of biomass from the compartment from elimination (EL), which is related to the mean characteristic turnover time of the organisms constituting the functional group of the given compartment (T_{PU} in weeks) and the consumption (= CON, i.e., the predation or grazing) by a secondary unit (SU; e.g., zooplankton).

$$dBM_{PU} = IPR_{PU} - CON_{PUSU} - EL_{PU} \quad (5.34)$$

For further details about the structure of the LakeWeb model, see section 8.5.

5.3.2.1. Phytoplankton

Fig. 5.12 gives the results for phytoplankton as this may be expressed first (A) by chlorophyll-a concentrations and by (B) algal volume. One can note that for the initial period, the modelled chlorophyll values are in good agreement with measured maximum and minimum values for the growing season (mainly data from July and August). Had the lake been shallow (curve 3), primary phytoplankton production and biomass would have been very high. Fig. 5.12B gives the "critical" algal volume of 5 mm^3/l and the "alarm" value of 10 mm^3/l. These limits are used by environmental authorities as practical guideline values (see Persson and Olsson, 1994). From fig. 5.12B one can see that had the lake been shallow, alarm values would dominate during summer, signaling increased risks of blooming of toxic cyanobacteria (blue-greens). In contrast, one can see that had the lake been large or deep (curves 5 and 4, respectively), the algal volumes would likely be well below the critical limit.

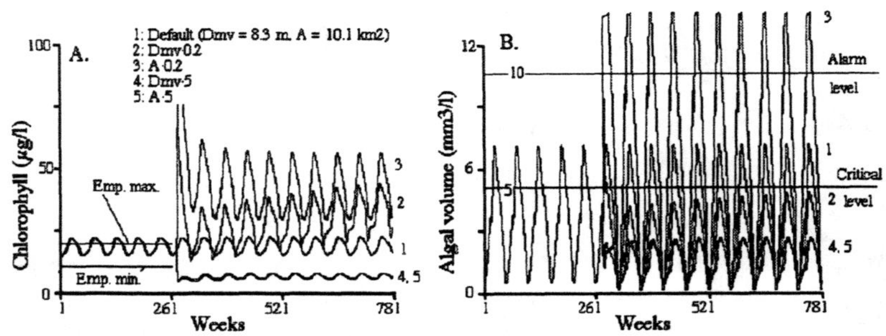

Fig. 5.12. Changes in chlorophyll-a concentrations (A) and algal volumes (B) from the changes in lake morphometry (mean depth set to 8.3 m, 0.2·8.3 = 1.7 m and 5·8.3 = 41.5 m; lake area set to 10.1 km^2 and to 0.2·10.1 = 2.02 and 5·10.1 = 50.5 km^2; basic data from Lake S. Bullaren, Sweden). For the initial period (weeks 1 to 261), the modelled values are also compared to the chlorophyll concentrations from the empirical model (the "norm"; see table 4.1); and the modelled values for algal volumes are compared to two operational management limits, the critical limit of 5 mm^3/l and the alarm limit of 10 mm^3/l.

5.3.2.2. Benthic algae

The corresponding results for benthic algae are given in fig. 5.13A and for bacterioplankton in fig. 5.13B. From fig. 5.13A, it is clear that for the initial period, the modelled values of the biomass of benthic algae correspond well with the empirical norm values (from table 4.1). Had the lake been large (curve 5), the benthic algal biomass would have been very high, and vice versa had it been small (curve 3). The form of the lake is important for the production of benthic algae; and it is interesting to note that had the lake been deep (curve 4), there would have been less benthic algae. Also the clarity of the water is very important for the benthic algae; and one can see that had the lake been shallow, with much SPM and a low Secchi depth, then the biomass of the benthic algae (curve 2) would have been lower than the default conditions in Lake S. Bullaren (curve 1).

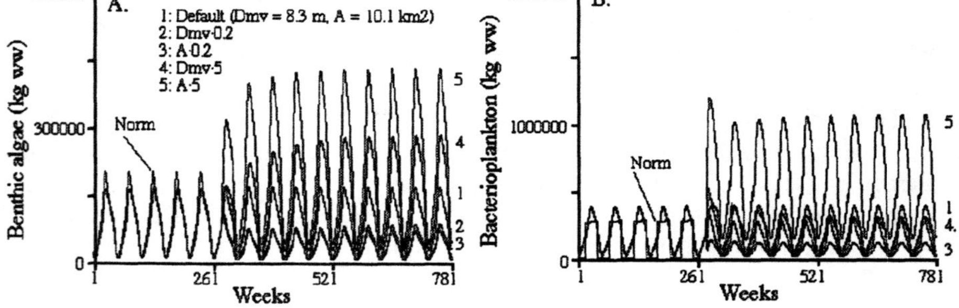

Fig. 5.13. Changes in biomass of benthic algae (A) and bacterioplankton biomass (B) from the changes in lake morphometry (mean depth set to 8.3 m, 0.2·8.3 = 1.7 m and 5·8.3 = 41.5 m; lake area set to 10.1 km^2 and to 0.2·10.1 = 2.02 and 5·10.1 = 50.5 km^2; basic data from Lake S. Bullaren, Sweden). For the initial period (weeks 1 to 261), the modelled values are also compared to the empirical norm values (see table 4.1).

5.3.2.3. Bacterioplankton

Fig. 5.13B gives the results for bacterioplankton biomass, and for the initial period the modelled values are in very good agreement with the norm values. Had the lake been large (curve 5), bacterioplankton production and biomass would have been high, and vice versa had the lake been small (curve 3). Bacterioplankton biomass reflects SPM very well since the main role of the bacterioplankton is to decompose suspended organic particles (the more SPM, the more bacterioplankton).

5.3.2.4. Macrophytes

The results for macrophyte cover (in % of the lake area) and macrophyte production are given in fig. 5.14A and B. The macrophyte cover is important since it influences man's access to the shoreline as well as the fish production. For example, it provides possibilities for the prey fish to escape hunting predatory fish. For the initial period, the modelled values of macrophyte production are close to the empirical norm values (fig. 5.14B). Had the lake been large (curve 5), the macrophyte cover and production would have been much higher than in the real lake (curve 1), and vice versa had the lake been small (curve 3). The form of the lake is important for the production of both macrophytes and benthic algae. One can note that had the lake been deep (curve 4), the macrophyte cover would have been very small. The Secchi depth is also important for the macrophyte production. And had the lake been deep (curve 4) with steep inclining walls, the production of macrophytes would have been lower than the default conditions (curve 1).

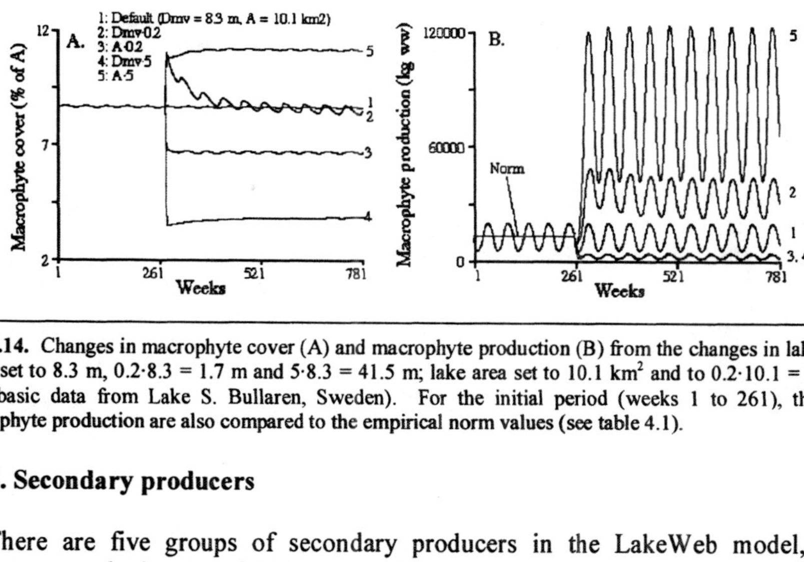

Fig. 5.14. Changes in macrophyte cover (A) and macrophyte production (B) from the changes in lake morphometry (mean depth set to 8.3 m, 0.2·8.3 = 1.7 m and 5·8.3 = 41.5 m; lake area set to 10.1 km^2 and to 0.2·10.1 = 2.02 and 5·10.1 = 50.5 km^2; basic data from Lake S. Bullaren, Sweden). For the initial period (weeks 1 to 261), the modelled values of macrophyte production are also compared to the empirical norm values (see table 4.1).

5.3.3. Secondary producers

There are five groups of secondary producers in the LakeWeb model, herbivorous and predatory zooplankton, zoobenthos, prey fish and predatory fish.

5.3.3.1. Zooplankton

Fig. 5.14 gives the results for the biomasses of herbivorous zooplankton (A) and predatory zooplankton (B). For the initial period, there is a good agreement between modelled values and norm values for herbivorous zooplankton and an acceptable agreement for predatory zooplankton. Note that these two empirical norms do not give seasonal variations. A large lake (curve 5) would have had much higher biomasses of these two groups of zooplankton. Note the interesting compensatory effects. There would be more bacterioplankton for the herbivorous zooplankton in the lake (fig. 5.13B), but less phytoplankton (fig. 5.12B). The net effect for the herbivorous zooplankton can be seen in fig. 5.14A. The biomass also depends on the grazing pressure; and one can note that in the large lake (curve 5), there would be much more predatory zooplankton grazing on the herbivorous zooplankton, which will keep this biomass low. It is also important to look at the scales of the y-axes and the order of the curves. For the deep lake (curve 4), the next highest biomass appears for herbivorous zooplankton and the next lowest biomass for predatory zooplankton. This result is mainly governed by the grazing pressure. Since there are relatively few predatory zooplankton in the deep lake, the predation pressure on the herbivorous zooplankton is relatively low, so the biomass can be high.

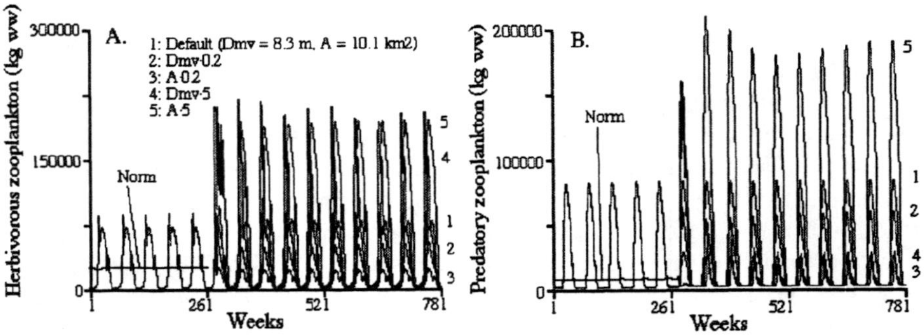

Fig. 5.15. Changes in biomass of herbivorous zooplankton (A) and biomass of predatory zooplankton (B) from the changes in lake morphometry (mean depth set to 8.3 m, 0.2·8.3 = 1.7 m and 5·8.3 = 41.5 m; lake area set to 10.1 km^2 and to 0.2·10.1 = 2.02 and 5·10.1 = 50.5 km^2; basic data from Lake S. Bullaren, Sweden). For the initial period (weeks 1 to 261), the modelled values are also compared to the empirical norm values (see table 4.1).

5.3.3.2. Zoobenthos

The results for zoobenthos are shown in fig. 5.16. Zoobenthos mainly eat "sediments" and to a lesser degree benthic algae. Zoobenthos are important food for the prey fish. For the initial period, the modelled values are in good agreement with the norm values. Deep and small lakes (curves 4 and 3) would likely have much less zoobenthos than large and shallow lakes (curves 5 and 2). Basically, zoobenthos production and biomass depend on sediment redox conditions and access to food.

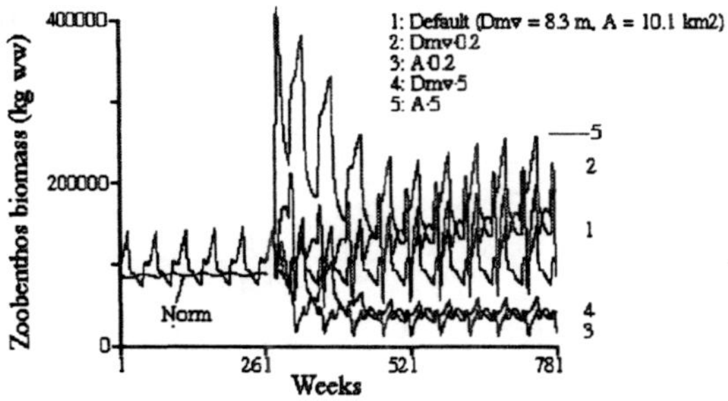

Fig. 5.16. Changes in zoobenthos biomass from the changes in lake morphometry (mean depth set to 8.3 m, 0.2·8.3 = 1.7 m and 5·8.3 = 41.5 m; lake area set to 10.1 km^2 and to 0.2·10.1 = 2.02 and 5·10.1 = 50.5 km^2; basic data from Lake S. Bullaren, Sweden). For the initial period (weeks 1 to 261), the modelled values are compared to the empirical norm values (see table 4.1).

5.3.3.3. Fish

Fish are evidently a very important part of the lake foodweb. Fig. 5.17 gives the results for the biomasses of prey fish (A) and predatory fish (B). There is a good agreement between modelled values and norm values for both categories of fish for the initial period. It has long been known (see fig. 3.5) that lake morphometry plays a decisive roll for the fish production. In large lakes (curve 5), one should expect much higher biomasses of these two groups of fish. In this case also, there are interesting effects of grazing. One can note that in the shallow highly eutrophic lake (curve 2), there are relatively more prey fish than predatory fish; and if the biomass of the predator is low, the prey biomass will increase. The scales on the y-axes show that there are also far more prey fish than predatory fish in lakes, and the curves illustrate the dynamic response to a sudden (although unrealistic) change in lake morphometry. One can note that predatory fish with a long turnover time react much more slowly than the other organisms.

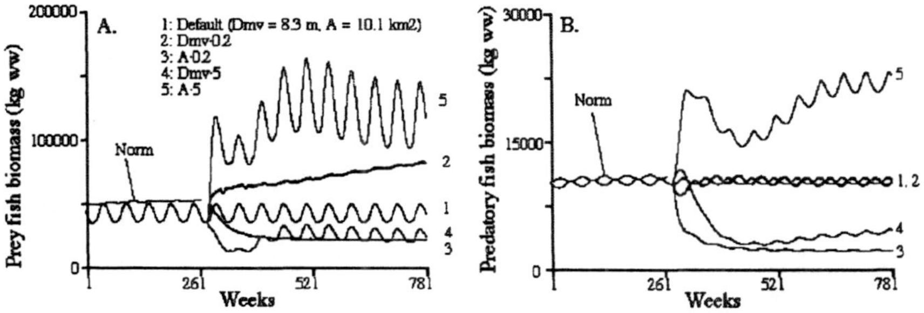

Fig. 5.17. Changes in prey fish biomass (A) and in biomass of predatory fish (B) from the changes in lake morphometry (mean depth set to 8.3 m, 0.2·8.3 = 1.7 m and 5·8.3 = 41.5 m; lake area set to 10.1 km^2 and to 0.2·10.1 = 2.02 and 5·10.1 = 50.5 km^2; basic data from Lake S. Bullaren, Sweden). For the initial period (weeks 1 to 261), the modelled values are compared to the empirical norm values (see table 4.1).

5.4. Conclusions

To summarize the results in chapter 5: Lake morphometry plays a decisive role in how a given lake would respond to a given nutrient loading. It is of paramount importance to understand this fact in contexts of lake management and science. If it is not properly understood, there can be no realistic predictions of the consequences of remedial measures or meaningful discussions about cause and effect of processes at the ecosystem scale.

6. Summary and Conclusions

Many people probably still believe in the view conveyed by most textbooks on lakes that information on lake size and form is of interest "here and there" and mainly for descriptive purposes. This book takes a holistic ecosystem perspective and focuses on lake structure and function. From that perspective, this book demonstrates a totally different view. Lake morphometry influences almost all transport processes in lakes, such as sedimentation, resuspension, mixing, diffusion, burial and outflow. Therefore, lake morphometry regulates concentrations of contaminants in water and sediments, and hence also ecosystem effect related to such concentrations. These transport processes are general and apply to all substances (nutrients, metals, organics, radionuclides, etc.). So, lake morphometry also regulates the nutrient concentrations from nutrient loading to lakes, and hence also primary production, and secondary production of zooplankton, zoobenthos and fish. Such relationships have been discussed in this book, where the aim has been to try to clarify the role of lake morphometry in contexts of mass-balance calculations and lake foodweb interactions. It has been demonstrated by numerous examples for different types of lakes that these aspects are fundamental for a basic understanding of how lakes work.

From the analysis related to the "interpretational key" in this book, it is possible to quantify the role of lake morphometry for important lake variables and relate variations among lakes to differences in catchment area characteristics, climatological factors, measurement uncertainties and lake morphometry. Conservative substances are much less influenced by lake morphometry than reactive substances, that is, substances participating in lake processes. The interpretational key should be of great interest in comparative lake studies to gain an understanding why certain lakes react as they do to, for example, contaminating substances or remedial measures. This knowledge is fundamental for lake management, and the basic understanding is fundamental for lake science.

7. Literature References

Abrahamsson, O. and Håkanson, L. 1998. Modelling seasonal flow variability of European rivers. Ecol. Modelling, 114:49-58.

Ahl, T. 1975. Effects of man-induced and natural loading of phosphorus and nitrogen on the large Swedish lakes. Int. Ver. Theor. Angew. Limnol., 19.

Ahlgren, I., Frisk, T. and Kamp-Nielsen, L. 1988. Empirical and theoretical models of phosphorus loading, retention and concentration vs. lake trophic state. Hydrobiologia, 170:285-303.

Aiken, G.R., McKnight, D.M., Werskaw, R.L. and MacCarthy, P. (eds) 1985. Humic Substances in Soil, Sediment and Water. New York: Wiley Interscience.

Allen, J.R.L. 1970. Physical Processes of Sedimentation. London: Allen and Unwin.

Ambio, 1976. Special issue on acid rain. Vol. 5. No. 5-6.

Baccini, P., Grieder, E., Stierli, R. and Goldberg, S. 1982. The influence of natural organic matter on the adsorption properties of mineral particles in lake water. Schweiz. Z. Hydrol., 44:99-116.

Beach Erosion Board, 1972. Waves in inland reservoirs. Tech. Mem. 132, Washington, D.C.: Beach Erosion Corps of Engineers.

Beeton, A.M., Edmondson, W.T. 1972. The eutrophication problem. J. Fish. Res. Canada, 29:673-682.

Benner, R., Moran, M.A. and Hodson, R.E. 1986. Biogeochemical cycling of lignocellulosic carbon in marine and freshwater ecosystems: relative contributions of proaryotes and eucaryotes. Limnol. Oceanogr., 31:89-100.

Bierman, V.J. Jr. 1980. A comparison of models developed for phosphorus management in the Great Lakes. In: Loehr, C., Martin, C.S. and Rast, W. (eds), Phosphorus Management Strategies for Lakes. Ann Arbor: Ann Arbor Science Publishers, pp. 235-255.

Bloesch, J. and Burns, N.M. 1980. A critical review of sedimentation trap technique. Schweiz. Z. Hydrol., 42:15-55.

Bloesch, J. and Uehlinger, U. 1986. Horizontal sedimentation differences in a eutrophic Swiss lake. Limnol. Oceanogr., 31:1094-1109.

Boers, P.C.M., Cappenberg, Th.E. and van Raaphorst, W. (eds) 1993. Proceeding of the Third International Workshop on Phosphorus in Sediments. Hydrobiologia, 253.

Boulion, V.V. 1994. Regularities of the primary production in limnetic ecosystems. St. Petersburg (in Russian).

Boulion, V.V. 1997. General characterization of some lakes in southern Karelia differing in the acidity and humic state. The response of lake ecosystems to changes in biotic and abiotic conditions. St. Petersburg, pp. 5-28 (in Russian).

Boulion, V.V. 2001. Contribution of macrophytes and phytobenthos in primary production of lake ecosystems. Kaliningrad: VIII Congress of Hydrobiological Society of Russian Academy of Sciences, pp. 158-159 (in Russian).

Bowen, H.J.M. 1966. Trace Elements in Biochemistry. London: Academic Press.

Boynton, W.R., Kemp, W.M. and Keefe, C.W. 1982. A comparative analysis of nutrients and other factors influencing estuarine phytoplankton production. In: Kennedy, V.S. (ed), Estuarine Comparisons. London: Academic Press, pp. 69-90.

Box, G.E.P. and Cox, D.R. 1964. An analysis of transformations. J. Royal Stat. Soc., Ser. B, 26:211-252.

Brinkhurst, R.O. 1974. The Benthos of Lakes. London: Macmillan.

Brittain, J.E. and Brabrand, Å. 2001. Fish movement in rivers, lakes and estuaries in relation to contamination by radionuclides. Freshwater Ecology & Inland Fisheries Laboratory (LFI), Natural History Museums, Norway: University of Oslo.

Brittain, J., Håkanson, L., Bergström, U. and Bjørnstad, H. E. 1994. The significance of hydrological and catchment processes for the transport and biological uptake of radionuclides in northern aquatic ecosystems. Proceedings of the 10th International Northern Research Basins Symposium and Workshop, Spitsbergen, Norway.

Burrough, P.A. 1986. Principles of Geographical Information Systems for Land Resources Assessment. Oxford: Clarendon Press.

Busch, W.D.N. and Sly, P.G. 1992. The Development of an Aquatic Habitat Cassification System for Lakes. Boca Raton, Florida: CRC Press.

Carlson, R.E. 1977. A trophic state index for lakes. Limnol. Oceanogr., 22:361-369.

Carlson, R.E. 1980. More complications in the chlorophyll - Secchi disk relationship. Limnol. Oceanogr., 25:379-382.

Carlsson, S. 1978. A model for the turnover of Cs-137 and potassium in pike (Esox Lucius). Health Phys., 35:549-554.

Chapra, S.C. 1980. Application of the phosphorus loading concept to the Great Lakes. In: Loehr, C., Martin, C.S. and Rast, W. (eds), Phosphorus Management Strategies for Lakes. Ann Arbor: Ann Arbor Science Publishers, pp. 135-152.

Chow, V.T. 1988. Applied Hydrology. New York: McGraw-Hill, Inc.

Christman, R.F. and Gjessing, E.T. (eds), 1983. Aquatic and Terrestrial Humic Materials. Ann Arbor: Ann Arbor Science Publishers.

Cornett, R.J. and Rigler, F.H. 1979. Hypolimnetic oxygen deficits: their prediction and interpretation. Science, 205:580-581.

Cummings, K.W. 1973. Trophic relations in aquatic insects. Ann. Rev. Entomol., 18:183-206.

Dahlgaard, H. (ed) 1994. Nordic radioecology, the transfer of radionuclides through nordic ecosystems to man. Amsterdam: Elsevier Science.

Draper, N.R. and Smith, H. 1966. Applied Regression Analysis. New York: John Wiley.

Duarte, C.M and Kalff, J. 1986. Littoral slope as a predictor of the maximum biomass of submerged macrophyte communities. Limnol. Oceanogr., 31:1072-1080.

Dubko, N.V. 1985. The labile and stable organic matter. Ecological system of Naroch lakes. Minsk, pp. 233-237 (in Russian).

Eberly, W.R. 1964. Further studies on the metalimnetic oxygen maximum, with special reference to its occurrence throughout the world. Investigations of Indiana Lakes and Streams, Vol. 6.

Einstein, H.A. 1950. The bed-load function for sediment transportation in open-channel flowes. U.S. Dept. Agric. Soil. Cons. Serv., T.B., No. 1026.

Engstrom, D.R. 1987. Influence of vegetation and hydrology on the humus budgets of Labrador lakes. Can. J. Fish. Aquat. Sci., 44:1306-1314.

Evans, R.D. and Håkanson, L. 1992. Measurement and prediction of sedimentation in small Swedish lakes. Hydrobiologia, 75:143-152.

Floderus, S. and Håkanson, L. 1989. Resuspension, ephemeral mud blankets and nitrogen cycling in Laholmsbukten, south east Kattegatt. Hydrobiologia, 176/177:61-75.

Förstner, U. and Wittmann, G.T.W. 1981. Metal Pollution in the Aquatic Environment. Berlin: Springer.

Galvenius, G. 1975. Bestämning av sjöbottnars brutenhet ur ekogram. Stockholm: Stencil, Tekniska Högskolan.

Gilbert, R. 1975. Sedimentation in Lillolet Lake, British Columbia. Can. J. Earth Sci., Vol. 12, No. 10.

Gjessing, E.T. 1976. Physical and Chemical Characteristics of Aquatic Humus. Ann Arbor: Ann Arbor Science Publishers.

Golterman, H.L. 1975. Physiological Limnology. Amsterdam: Elsevier.

Gorham, E. Dean, W.E. and Sanger, J.E., 1983. The chemical composition of lakes in the north-central United States. Limnol. Oceanogr., 28:287-301.

Gorham, E. Underwood, J.K., Martin, F.B. and Ogden III, J.B. 1986. Natural and anthropogenic causes of lake acidification in Nova Scotia. Nature, 324:451-453.

Håkanson, L. 1974. A mathematical model for establishing numerical values of topographical roughness for lake bottoms. Geog. Ann., 56:183-200.

Håkanson, L. 1977. The influence of wind, fetch, and water depth on the distribution of sediments in Lake Vänern, Sweden. Can. J. Earth Sci., 14:397-412.

Håkanson, L. 1978a. Optimization of lake hydrographic surveys. Wat. Res. Res., 14:545-560

Håkanson, L. 1978b. The length of closed geomorphic lines. Mathematical Geology, 10:141-167.

Håkanson, L. 1981a. A Manual of Lake Morphometry. New York: Springer-Verlag.

Håkanson, L. 1981b. Determination of characteristic values for physical and chemical lake sediment parameters. Water Resources Research, 17:1625-1640.

Håkanson, L. 1991a. Ecometric and Dynamic Modelling - Exemplified by Cesium in Lakes After Chernobyl. Berlin: Springer-Verlag.

Håkanson, L. 1991b. Sediment variability. Burton, Jr., A.G., (ed). Contaminated Sediment Toxicity Assessment. Boca Raton, Florida: Lewis Publishers, pp. 19-35.

Håkanson, L. 1995a. Models to predict organic content of lake sediments. Ecol. Modelling, 82:233-245.

Håkanson, L. 1995b. Fiskodling och miljöeffekter i sjöar - nya resultat motiverar nya bedömningsunderlag. Vatten 51, Lund, Bloms Boktryckeri.

Håkanson, L. 1999. Water Pollution - Methods and Criteria to Rank, Model and Remediate Chemical Threats to Aquatic Ecosystems. Leiden: Backhuys Publishers.

Håkanson, L. 2000. Modelling Radiocesium in Lakes and Coastal Areas - New Approaches for Ecosystem Modellers. A Textbook With Internet Support. Dordrecht: Kluwer Academic Publishers.

Håkanson, L. 2003a. Quantifying burial, the transport of matter from the lake biosphere to the geosphere. Internat. Rev. Hydrobiol., 88:539-560.

Håkanson, L. 2003b. A general management model to optimize lake liming operations. Lakes & Reservoirs: Research & Management, 8:105-140.

Håkanson, L. and Boulion, V. 2002. The Lake Foodweb - Modelling Predation and Abiotic/Biotic Interactions. Leiden: Backhuys Publishers.

Håkanson, L. and Jansson, M. 1983. Principles of Lake Sedimentology. Berlin: Springer.

Håkanson, L. and Johansson, H. 2003. Models to predict the distribution coefficient (dissolved and particulate) for phosphorus in lakes. Manuscript, Institute. of Earth Sciences, Uppsala University.

Håkanson, L. and Lindström, M. 1997. Frequency distributions and transformations of lake variables, catchment area and morphometric parameters in predictive regression models for small glacial lakes. Ecol. Modelling, 99:171-201.

Håkanson, L., Parparov, A. and Hambright, K.D. 2000. Modelling the impact of water level fluctuations on water quality (suspended particulate matter) in Lake Kinneret, Israel. – Ecol. Modelling, 128:101-125.

Håkanson, L. and Peters, R.H. 1995. Predictive Limnology. Methods for Predictive Modelling. Amsterdam: SPB Academic Publishing.

Håkanson, L., and Sazykina, T. 2001. A blind test of the MOIRA lake model for radiocesium for Lake Uruskul, Russia, contaminated by fallout from the Kyshtym accident in 1957. J. Env. Radioactivity, 54:327-344.

Håkanson, L., Sazykina, T.G., and Kryshev, I.I. 2002. A general approach to transform a lake model for one radionuclide (radiocesium) to another (radiostrontium) and critical model tests using data for four Ural lakes contaminated by the fallout from the Kyshtym accident in 1957. J. Env. Radioactivity, 60:319-350.

Håkanson, L., Blenckner, T. and Malmaeus, J.M. 2003. Defining new, general methods to define surface water and deep water, outflow and internal loading for mass-balance model for lakes. Ecol. Modelling (in press).

Hansen, K. 1961. Lake types and lake sediments. Verh. Int. Ver. Limnol., 14:285-290.

Harden-Jones, F.R. 1968. Fish Migration. London: Edward Arnold Publishers.

Hayes, M.H.B., MacCarthy, P., Malcolm, R.L. and Swift, R.S. (eds), 1989. Humic Substances II. In Search of Structure. New York: Wiley Interscience.

Hedenstierna, B. 1948. Stockholms skärgård. Geografiska annaler, Häfte 1-2.

Hellström, T. 1991. The effect of resuspension on algal production in a shallow lake. Hydrobiologia 213:183-190.

Hilton, J. 1985. A conceptual framework for predicting the occurrence of sediment focusing and sediment redistribution in small lakes. Limnol. Oceanogr., 30:1131-1143.

Horne, A.J. and Goldman, C.R. 1994. Limnology. New York: McGraw-Hill.

Hutchinson, G.E. 1957. A Treatise on Limnology. Vol. 2. Geography, Physics and Chemistry. London: Wiley.

Hutchinson, G.E. 1967. A Treatise on Limnology II: Introduction to Lake Biology and the Limnoplankton. New York: Wiley.

IAEA, 1988. International Atomic Energy Agency. Assessing the impact of deep sea disposal of low level radioactive waste on living marine resources. Tech. Rep. No. 288, Vienna.

IAEA, 2000. Modelling of the transfer of radiocaesium from deposition to lake ecosystems. Report of the VAMP Aquatic Working Group. International Atomic Energy Agency, Vienna: IAEA-TECDOC-1143.

Ivanova, M.B. 1997. The impact of active reaction and total mineralization of water on the formation of zooplankton community in lakes with nearly extreme values of these factors. The response of lake ecosystems to changes in biotic and abiotic conditions. St. Petersburg, pp. 71-86 (in Russian).

Ivanova, M.B. Boulion, V.V., Nikulina, V.N., Paveljeva, E.B., Ilyashuk, B.P., Polyakova, E.A., Anokhina, L.E., 1993. Limnological characteristics of natural acidic lakes in the north-west of Russia. Russian Journal of Aquatic Ecology, 2:81-90.

Jackson, T.A. and Hecky, R.E. 1980. Depression of primary productivity by humic matter in lake and reservoir waters of the boreal forest zone. Can. J. Fish. Aquat. Sci., 37:2301-2317.

Janus, L.L. and Vollenweider, R.A. 1981. The OECD cooperative report on eutrophication: Canadian contribution. Summary Rep. Sci. Ser. No. 132.

Jimenez, F. and Gallego, E. 1998. Effects of radiation on aquatic organisms. Preprint, from the MOIRA project, Univ. de Politech. de Madrid, Spain.

Jonsson, A. 1997. Whole lake metabolism of allochthonous organic material and the limiting nutrient concept in Lake Örträsket, a large humic lake in northern Sweden. Thesis, Sweden: Umeå University.

Jørgensen, S.E. and Johnsen, J. 1989. Principles of Environmental Science and Technology. Studies in Environmental Science, 33. Amsterdam: Elsevier.

Kalff, J. 2002. Limnology: Inland Water Ecosystems. Upper Saddle River, New Jersey: Prentice Hall.

Khailov, K. M. (ed), 1974. The biochemical trophodynamics in coastal sea ecosystems. "Naukova Dumka", Kiev, 176 (in Russian).

Kiefer, D.A. and Austin, R.W. 1974. The effect of varying phytoplankton concentration on submarine transmission in the Gulf of California. Limnol. Oceanogr., 19:55-64.

Kirk, J.T.O. 1983. Light and Photosynthesis in Aquatic Ecosystems. Cambridge: Cambridge University Press.

Konitzer, K. and Meili, M. 1997. Redistribution of sedimentary Cs-137 in small Swedish lakes after the Chernobyl fallout 1986. In: Desmet, G. (et al., eds), Freshwater and estuarine radioecology, pp. 167-172, Amsterdam: Elsevier.

Konoplev, A., Bulgakov, A., Hilton, J., Comans, R. and Popov, V. 1997. Long-term kinetics of radiocesium fixation by soils. In: Desmet, G. (et al., eds), Freshwater and estuarine radioecology, pp. 173-182, Amsterdam: Elsevier.

Kranck, K. 1973. Flocculation of suspended sediment in the sea. Nature, 246:348-350.

Kranck, K. 1979. Particle matter grain-size characteristics and flocculation in a partially mixed estuary. Sedimentology, 28:107-114.

Kristensen, P., Søndergaard, M. and Jeppesen, E. 1992. Resuspension in a shallow eutrophic lake. Hydrobiologia, 228:101-109.

Krylov, P.I., Plyakova, E.I., Galimov, Y.R. 1997. Zooplankton of an acidic lakes: Strategies for survival under conditions of food limitation. The response of lake ecosystems to changes in biotic and abiotic conditions. St. Petersburg, pp. 87-106 (in Russian).

Lehtinen, K-J., Gyllenhammar, A., Håkanson, L., Mattsson, K., Engström, C., Tana, J. 2001. Fiskodling i kassar som en hållbar näringsgren i skärgården. Interreg IIa (71/97). Sammanfattande slutrapport.

Lick, W., Lick, J. and Ziegler, C.K. 1992. Flocculation and its effect on the vertical transport of fine-grained sediments. Hydrobiologia, 235/236:1-16.

Likens, G., Wright, R., Galloway, J. and Butler, T. 1979. Acid rain. Sci. Am., 241:43-51.

Lindqvist, O., Johansson, K., Aastrup, M., Andersson, A., Bringmark, L., Hovsenius, G., Håkanson, L., Iverfeldt, Å., Meili, M. and Timm, B. 1991. Mercury in the Swedish environment. Water, Air and Soil Pollution, Vol. 55.

Lindström, M. 2000. Predictive modelling of heavy metals in urban lakes. Dr. thesis, Sweden: Uppsala University.

Lindström, M., Håkanson, L., Abrahamsson, O. and Johansson, H. 1999. An empirical model for prediction of lake water suspended matter. Ecol. Modelling, 121:185-198.

Loeb. S.L., Reuter, J.E. and Goldman, C.R. 1983. Littoral zone production of oligotrophic lakes. The contributions of phytoplankton and periphyton. Periphyton of freshwater ecosystems. Dr. W. Junk Publisher, The Hague, pp. 161-167.

Madruga, M.J. and Cremers, A. 1997. On the differential binding mechanisms of radiostrontium and radiocesium in sediments. In: Desmet, G. (et al., eds), Freshwater and Estuarine Radioecology, pp. 207-216, Amsterdam: Elsevier.

Mann, K.H. 1982. Ecology of Coastal Waters. A Systems Approach. London: Blackwell Scientific Publications.

McDowall, R. M. 1988. Diadromy in Fishes. Portland, Oregon: Timber Press.

Meeuwig, J.J. and Peters, R.H. 1996. Circumventing phosphorus in lake management: a comparison of chlorophyll-a predictions from land-use and phosphorus-loading models. Can. J. Fish. Aq. Sci., 53:1795.

Meili, M. 1991a. In situ assessment of trophic levels and transfer rates in aquatic food webs using chronic (Hg) and pulsed (Chernobyl Cs-137) environmental contaminants. Verh. Int. Verein. Limnol., 24.

Meili, M. 1991b. Mercury in boreal forest lake ecosystems. Acta Univ. Upsaliensis 336. Thesis, Uppsala University, Sweden.

Merilehto, K., Kenttämies, K. and Kämäri, J. 1988. Surface water acidification in the ECE region. Nordic Council of Ministers, NORD 1988:89.

Moberg, L. (ed), 1991. The Chernobyl fallout in Sweden. The Swed. Rad. Prot. Inst., Stockholm, 633 p.

Monitor, 1991. Acidification and liming of Swedish freshwaters. Swedish Environmental Protection Agency, Solna.

Mortimer, C.H. 1952. Water movements in lakes during summer stratification; evidence from the distribution of temperature in Windermere. Phil. Trans. Royal Soc. London, Ser. B, boil. Sciences, No. 635, Vol. 236.

Mosteller, F. and Tukey, J.W. 1977. Data Analysis and Regression: A Second Course in Statistics. Reading, Massachusetts: Addison-Wesley Publishing.

Newman, M.C. 1993. Regression analysis of log-transformed data: statistical bias and its correction. Env. Tox. and Chem., 12:1129-1133.

Nõges, P., Tuvikene, L., Nõges, T. and Kisand, A. 1999. Primary production, sedimentation and resuspension in large shallow Lake Võrtsjärv. Aquatic Sciences, 61:161-182.

Norrman, J.O. 1964. Lake Vättern. Investigations on shore and bottom morphology. Geogr. Ann., Häfte 1-2.

Northcote, T.G. 1978. Migratory strategies and production in freshwater fishes. In: Gerking, S.D. (ed). Ecology of Freshwater Fish Production. Oxford: Blackwell Scientific Publications, pp. 326-359.

Nürnberg, G.K. and Shaw, M. 1998. Productivity of clear and humic lakes: nutrients, phytoplankton, bacteria. Hydrobiologia, 382:97-112.

OECD, 1982. Eutrophication of Waters. Monitoring, Assessment and Control. Paris: Organization for Economic Co-operation and Development.

Ostapenia, A.P. 1985. Ratio between the components of seston. Ecological system of Naroch lakes. Minsk, pp. 232-233 (in Russian).

Ostapenia, A.P. 1987. Ratio between particulate and dissolved organic matter in waters of different types. Production-hydrobiological investigations of water ecosystems. Leningrad, pp. 109-115 (in Russian).

Ostapenia, A.P. 1989. Seston and detritus as structural and functional components of water ecosystems. Thesis for a Doctor's degree. Kiev (in Russian).

Ostapenia, A.P., Pavljutin, A.P. and Zhukova, T.V. 1985. Conditions and factors determining quality of waters and trophic status of lakes. Ecological system of Naroch lakes. Minsk, pp. 263-269 (in Russian).

Ottosson, F., and Abrahamsson, O. 1998. Presentation and analysis of a model simulating epilimnetic and hypolimnetic temperatures in lakes. Ecol. Modelling, 110:223-253.

Overrein, L.N., Seip, H.M and Tollan, A. 1980. Acid precipitation - effect on forest and fish. Final report on the SNSF-project 1972-1980. Oslo-Ås.

Pearson, T.H. and Rosenberg, R. 1976. A comparative study on the effects on the marine environment of wastes from cellulose industries in Scotland and Sweden. Ambio, 5:77-79.

Persson, G. and Olsson, H. 1994. Eutrophication of Swedish lakes and rivers - present status, development, causes and consequences (in Swedish, Eutrofiering i svenska sjöar och vattendrag - tillstånd, utveckling, orsak och verkan). Swedish Environmental Protection Agency, Report 4147, Stockholm.

Peters, R.H. 1986. The role of prediction in limnology. Limnol. Oceanogr., 31:1143-1159.

Peters, R.H. 1991. A critique for ecology. Cambridge: Cambridge University Press.

Peterson, R.C. Jr. 1991. The contradictory biological behavior of humic substances in aquatic environment. pp. 369-390. In: Allard, B., Borén, H. and Grimvall, A. (eds), 1991. Humic Substances in Aquatic and Terrestrial Environment. Heidelberg: Springer.

Pfaffenberger, R.C. and Patterson, J.H. 1987. Statistical Methods for Business and Economics. Homewood, Illinois: Irwin.

Pilesjö, P., Persson, J. and Håkanson, L. 1991. Digital bathymetric information for calculations of morphometrical parameters and surface water retention time for coastal areas (in Swedish). National Swedish Environmental Protection Agency (SNV) Report no. 3916, Solna, Sweden.

Pokrovskaja, T.N., Mironova, N.Ja. and Shilkrot, G.S. 1983. Macrophyte lakes and their eutrophication. Moscow, Nauka (in Russian).

Prairie, Y. 1996. Evaluating the predictive power of regression models. Can. J. Fish. Aquat. Sci., 53:490-492.

Preisendorfer, R.W. 1986. Secchi disk science: Visual optics of natural waters. Limnol. Oceanogr., 31:909-926.

Rasmussen, J.B. and Kalff, J. 1987. Empirical models for zoobenthic biomass in lakes. Can. J. Fish. Aquat. Sci., 44:990-1001.

Rasmussen, J.B., Godbout, L. and Schellenberg, M. 1989. The humic content of lake water and its relationship to watershed and lake morphometry. Limnol. Oceanogr., 34:1336-1343.

Rawson, D.S. 1955. Morphometry as a dominant factor in the productivity of large lakes. Verh. Int. Ver. Limnol. 12:164-175.

Rosenberg, R. 1985. Eutrophication - the future marine coastal nuisance? Mar. Pollut. Bull., 16:227-231.

Salomons, W. and Förstner, U. 1984. Metals in the Hydrocycle. Heidelberg: Springer.

Scheffer, M. 1990. Multiplicity of stable states in freshwater systems. Hydrobiologia, 200/201:475-486.

Scheffer, M., Brock, W. and Westley, F. 2000. Socioeconomic mechanisms preventing optimum use of ecosystem services: An interdisciplinary theoretical analysis. Ecosystems, 3:451-4571.

Schindler, D.W. 1971. A hypothesis to explain differences and similarities among lakes in the Experimental Lakes Area, northwestern Ontario. J. Fish. Res. Bd. Can., 28:295-301.

Schindler, D.W. 1977. Evolution of phosphorus limitation in lakes. Science, 195:260-262.

Schindler, D.W. 1978. Factors regulating phytoplankton production and standing crop in the world's freshwaters. Limnol. Oceanogr., 23:478-486.

Shadrin, N.V. 1985. The dependence of production characteristics on morphometric parameters of the water body. In: The hydrobiology and hydroparasitology of Near Bajkal region, Novosibirsk, Nauka, pp. 201-205 (in Russian).

Shadrin, N.V. 2003. Is it possible to quantitatively assess the role of algobacterial films in a water body? In: Krumbein, W.G., Paterson, D.M. and Zavarzin, G.A. (eds), 2003. Fossil and Recent Biofilms - A Natural History of Life on Earth. Dordrech, The Netherlands: Kluwer Academic Publishers, pp. 353-361.

Skidmore, A.K. 1990. Terrain position as mapped from a gridded digital elevation model. Int. Journal of Geographical Information Systems, 4: 33-49.

Smith, I.R. and Sinclair, I.J. 1972. Deep water waves in lakes. Freshwater Biol. 2: 387-399.

Spence, D.H.N. 1982. The zonation of plants in freshwater lakes. Adv. Ecol. Res., 12:37-125.

Stokes, G.G. 1851. Collected papers. Vol. III. Cambridge Trans. Vol. IX (see, e.g., Lamb, H. 1945. Hydrodynamics. New York: Gover Publishing).

Straskraba, M. 1980. The Effect of Physical Variables on Freshwater Production: Analyses Based on Models. The Functioning of Freshwater Ecosystems. London: Cambridge University Press, pp. 13-84.

Stumm, W. and Morgan, J.J. 1981. Aquatic Chemistry. New York: Wiley Interscience.

Sturm, M. 1975. Depositional and erosional sedimentary features in a turbidity current controlled basin (Lake Brienz). IX International Congress of Sedimentology, Nice, 1975, Theme 5, Vol. 5/2.

Taylor, J.K. 1990. Statistical Techniques for Data Analysis. Chelsea, Michigan: Lewis Publishers.

Thurman, E.M. 1985. Organic Geochemistry of Natural Waters. Dordrecht: Martinus Nijhoff/Dr. W. Junk Publishers.

Tilzer, M.M. 1988. Secchi disk - chlorophyll relationships in a lake with highly variable phytoplankton biomass. Hydrobiologia, 162:163-171.

Velimorov, B. 1991. Detritus and the concept of non-predatory loss. Arch. Hydrobiol., 121:1-20.
Vollenweider, R.A. 1958. Sichttiefe und Production. Verh. Int. Ver. Limnol., 13:142-143.
Vollenweider, R.A. 1960. Beitrage zür Kenntnis optischer Eigenschaften der Gewässer und Primärproduktion. Mem. Ist. Ital. Idrobiol., 12:201-244.
Vollenweider, R.A. 1968. The scientific basis of lake eutrophication, with particular reference to phosphorus and nitrogen as eutrophication factors. Tech. Rep. DAS/DSI/68.27, Paris: Organization for Economic Co-operation and Development.
Vollenweider, R.A. 1976. Advances in defining critical loading levels for phosphorus in lake eutrophication. Mem. Ist. Ital. Idrobiol., 33:53-83.
Vorobev, G.A. 1977. Landscape types of lake overgrowing. Natural conditions and resources of North European part of the Soviet Union. Vologda, pp. 48-60 (in Russian).
Wallin, M., Håkanson, L. and Persson, J. 1992. Load Models for Nutrients in Coastal Areas, Especially from Fish Farms (in Swedish with English summary). Copenhagen: Nordiska ministerrådet, 1992:502.
Welch, P.S. 1948. Limnological Methods. Toronto: The Blakiston Company.
Wershaw, R.L., Burcar, P.J. and Goldberg, M.C. 1969. Interaction of pesticides with natural organic material. Env. Sci. Tech., 3:271-273.
Westlake, D.F. 1980. Primary production. The Functioning of Freshwater Ecosystems. London: Cambridge University Press, pp. 141-246.
Wetzel, R.G. 2001. Limnology. London: Academic Press.
Wetzel, R.G. and Likens, G.E. 1990. Limnological Analyses. Heidelberg: Springer.
Whicker, F.W. and Schultz, V. 1982. Radioecology: Nuclear Energy and the Environment: Volume 1. Boca Raton, Florida: CRC Press.
Winberg, G.G. 1985. Main features of production process in the Naroch lakes. Ecological system of Naroch lakes. Minsk, pp. 269-284 (in Russian).
Wotton, R. J. 1990. Ecology of Teleost Fishes. New York: Chapman and Hall.

8. Appendices

These appendices include information that may not be so vital to an understanding of the basic topics of this book. They provide more detailed information on regression analyses, on some of the lakes discussed in chapter 4, on models to predict water discharge and lake temperatures from standard map data and on the basic equations for the LakeMab and LakeWeb models.

8.1. Basic concepts in regression analysis

Regression analyses could be performed for many reasons, for example, to compare values predicted from models (generally x) with empirical data (y), to test hypotheses about relationships, and to develop statistical/empirical models. Many textbooks examine regression analyses (see Draper and Smith, 1966; Mosteller and Tukey, 1977; Pfaffenberger and Patterson, 1987; Newman, 1993). This appendix (basically from Håkanson and Peters, 1995) presents a brief summary of basic concepts that must be understood to use this textbook. Fig. 8.1 summarizes some central concepts in regression analyses that are of particular interest in lake studies.

Any regression is based on pairs of data (x_i, y_i). Generally, the x and the y samples should be normally distributed. If a normal distribution is not the case for the absolute values, different types of transformations (Box and Cox, 1964) can be used to obtain more compatible, and more normally distributed, x- and y-variables. Some common transformations are the logarithmic (log10 = log or ln = logn), or different exponentials like $\sqrt{x} = x^{0.5}$, x^2, x^{-1}, or $1/(x+\text{const})$. The data could also be ranked relative to the highest and/or lowest numerical value in the series. Certain transformations (like ex) maximize the weight of high values in regressions. Others, such as log(x), minimize the weight of high values. There are linear regressions ($y = a\cdot x + b$), non-linear regressions (for example, $y = a\cdot\log(x) + b$) and multiple regressions (like $y = a\cdot x_1 + b\cdot\log(x_2) + c\cdot 10x_3 + d$). The x and the y values may be either single values or functions (like $x = a\cdot z_1 + b/z_2$).

Two fundamental characteristics of any regression analysis are the number of data (n) used in the regression and the range of the x- and y-variables. Very different assumptions are implied in building predictive regression models if:

1. the empirical data base is very small ($n<10$), small ($n<30$) or large ($n>100$), and/or
2. the range in the x- and y-variable is small relative to the entire possible range of the variable.

There are several methods to establish a regression line. Generally, the distances (d), or rather the square of the distances, between the points on the regression line (x', y') and the data points (x_i, y_i) are minimized. The slope of the regression line is given by the x constant (a) in the regression equation $y = a\cdot x+b$. The intercept, b, is the y value when $x = 0$. Regression analysis can be a very useful method to find the line that best fits the data, often indicated by:

- The best-fit regression line between one x-variable and one y-variable (simple linear regression), or between several x-variables and one y-variable (for example, by stepwise multiple regression analysis).
- The correlation coefficient (r), or r^2 value. The r^2 is sometimes called the coefficient of determination or the degree of statistical explanation since perfect fit, 100% statistical explanation, exists when $r^2 = 1$, and no statistical explanation at all is obtained if $r^2 = 0$.

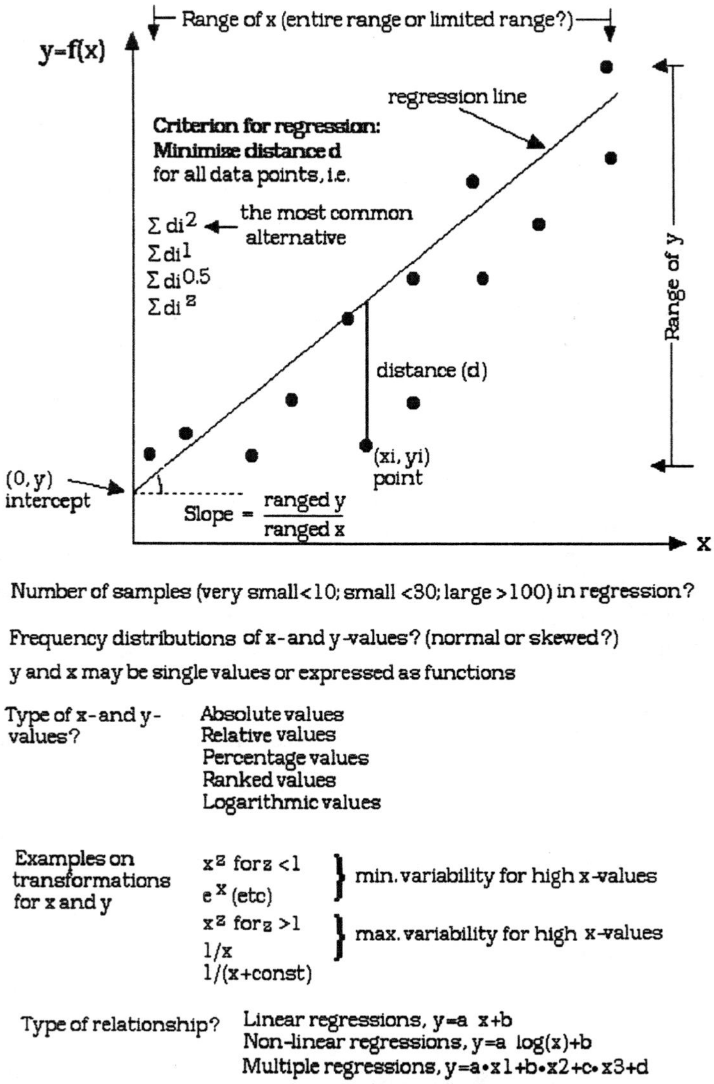

Fig. 8.1. A compilation of key concepts related to regression analyses (modified from Håkanson and Peters, 1995).

- The p level. The probability level gives the statistical significance of the regression. Usually, one would like to know if the variations in x are significantly linked to the variations in y. The strength of the relationship can be described by the correlation coefficient, r, or by the r^2 value. Usually in limnology, one would prefer to use the simple r^2 value (and not the adjusted r^2 value). The r^2 value is highly dependent on the number of data (n), the range of the data, the transformation, and the residual scatter. If the other characteristics are constant, the level of statistical significance (p) in the correlation or regression is related to both n and r^2. The statistical derivation of the r^2 value is left to more advanced statistical textbooks. A nomogram (fig. 8.2) relating the values of r^2, n and p, illustrates these relationships. If the x and y variables are normally distributed, then the p value could be read directly from this graph. The 95% confidence level (p = 0.05) has been marked in fig. 8.2. because this criterion is often used as a default value for significance, but the 90% and the 99% levels are also common.

- The variance and standard deviation (SD) of the sample. The variance is the average of the square of the deviations from the regression line, $\sum d_i^2/(n-1)$. The standard deviation is the square root of the variance; it may be considered as the "average error" because it is the average distance between the values of y predicted by the regression line and the observed values of y in the data set used to develop the regression. This statistic is fundamental to defining the 95% confidence limits around predicted y-values or to approximating them (as 2·SD). The standard deviation and variance are relatively independent of range and n, once the sample is above about 30.
- The F value. This value is used to test the significance of the regression. Many statistical software programs set F to 4 as a default value. At F = 4, it means that there is a high probability that a given x variable actually influences the y variable. If the F value is lower, random and nonsensical x variables are more likely to be included in the regression. For statistical F tests, see Taylor (1990).

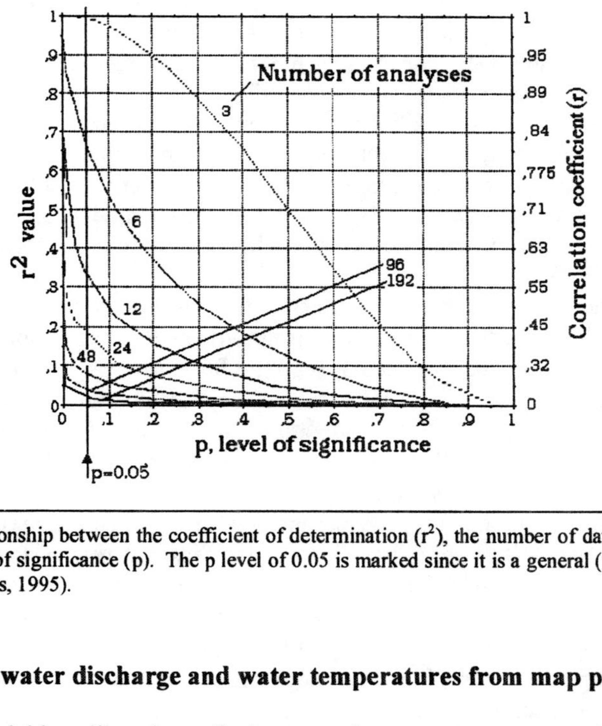

Fig. 8.2. The relationship between the coefficient of determination (r^2), the number of data pairs in the regression (n) and the statistical level of significance (p). The p level of 0.05 is marked since it is a general (95%) level of significance (from Håkanson and Peters, 1995).

8.2. Predicting water discharge and water temperatures from map parameters

Climatic variables, like river discharge and water temperature, influence many important processes in a lake ecosystem. For example, the retention time of the water and, hence, the retention of substances in lake water, are regulated to a large extent by the water discharge in the tributary rivers. Differences in hydrological regime also affect the transport of nutrients and toxic substances from catchments (Brittain et al., 1994).

The following models for water discharge (Q) and lake water temperatures can be run not using lake morphometric parameters but from related readily accessible map parameters, such as altitude, latitude and continentality (= distance to the ocean).

8.2.1. Water discharge

River discharge depends on many more or less stochastic processes and has a high degree of variability between years for a given river. It is difficult to give a reliable prediction of the water discharge (Q) for a specific river at a given time. A standard procedure is to measure the river discharge for a long period of time (decades) and then give a statistical estimate of a probability that Q is going to be within a certain range at a certain time. That method is appropriate for many purposes, providing that a sufficiently long and reliable set of empirical data is available (Chow, 1988). If empirical data on Q are not available, which is certainly the case for a very large number of rivers, other methods are necessary, such as statistical/empirical methods to predict Q from soil type distributions, vegetation types, etc. Such models (like HFAM by Hydrocomp Inc.) can be very precise and valuable, but they require a lot of costly (both in terms of time and money) field data and also site-specific river data for the calibration.

The model discussed here (see fig. 8.3) has been presented by Abrahamsson and Håkanson (1998) to meet specific demands in ecosystem modelling rather than in hydrology. The first requirement is that this model must be based on readily available driving variables, preferably from standard maps. There are many uncertainties in ecosystem models, but all uncertainties are not of equal importance for the predictive success of the model. There will always be uncertainties concerning the proper value for Q. This model is meant to yield predictions of Q that can be accepted in ecosystem models where (i) the focus is on, for example, the predictive power for the concentration of toxins in water, sediments and biota or (ii) in ecosystem modelling when the target variables to be predicted are biological variables (like fish biomasses) and/or chemical variables (like lake pH and phosphorus concentration). In such ecological contexts, the inevitable uncertainties in the predicted values of Q associated with this simple model can be accepted. This argument explains why this Q model is based on readily available map parameters, such as latitude, altitude and precipitation.

To calibrate and validate the Q model, an extensive data set from more than 200 European rivers were used. The discharge of the chosen rivers is not affected by regulation for hydropower or irrigation purposes since that produces unnatural seasonal flow patterns. The time series for the monthly data were at least six years long, and some as long as 80 years. The data sets were divided in two parts of equal size; one for the calibration and the other for the validation. From fig. 8.3, one can note that the only obligatory driving variables for this Q model are altitude, latitude, mean annual precipitation and catchment area.

To simulate the monthly variations in Q, six seasonal variability norms are utilized for Europe (see fig. 8.3). A seasonal variability norm is used to add a seasonal pattern to an annual value (Håkanson and Peters, 1995). Two of these norms should represent the typical seasonal flow pattern in the most southern and northern parts of Europe, respectively. Two other norms should describe the effect of altitude on monthly variability in Q. Depending on the location (latitude and altitude) and the mean annual discharge of the specific river, the six seasonal variability norms are weighted together and a site-specific seasonal variability norm is calculated. In the calibration, it became obvious that the addition of longitude did not significantly increase the degree of explanation of the model and longitude was, therefore, excluded.

To account quantitatively for how latitude (Lat), altitude (Alt), and mean annual discharge (Q_{mv}) for a specific river influence the seasonal (monthly) variability, different weighing factors ranging from zero to one were developed. For example, the weighting factor for latitude should be zero for a river in the southernmost part of Europe (35°N) and one in the northernmost part (70°N). To account for the fact that the relations do not have to be linear, each weighting factor was given an exponent. The exponents are used to control in which range the changes in the parameters are most critical. This approach is, of course, still a very simple way to simulate the influence of different parameters. This approach gave the equation for the seasonal moderator for Q (Y_Q) given in fig. 8.3. The values for the six norms are also given in fig. 8.3.

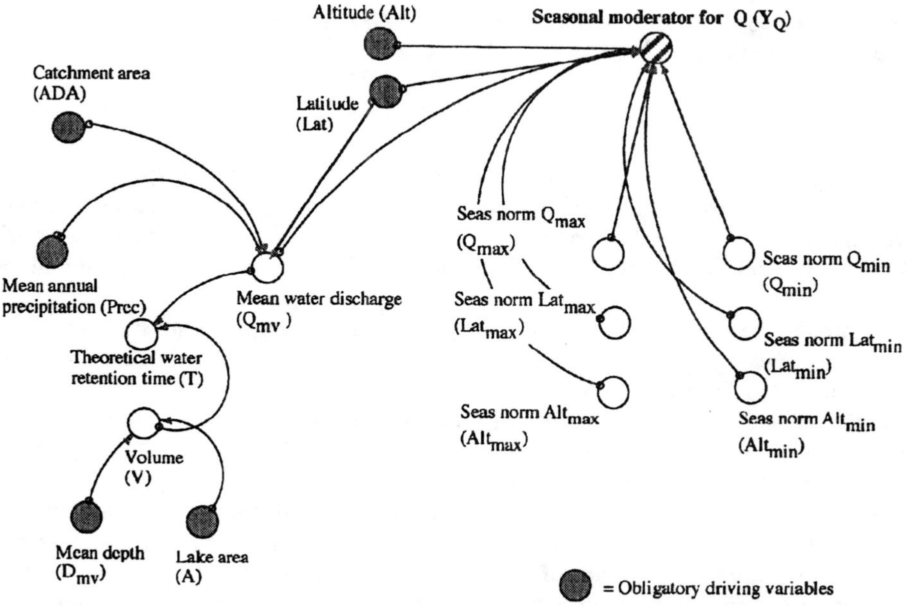

Equations:

Q_{mv} (m^3/s)= $0.01 \cdot ADA \cdot (10^{-6}) \cdot (Prec/650) \cdot ((Lat+1)/61)$

V (m^3) = $A \cdot D_{mv}$

T (months) = $V/(Q_{mv} \cdot 60 \cdot 60 \cdot 24 \cdot 30)$

$Y_Q = 1+0.526 \cdot ((Lat-35)^{2.18}/35^{2.18} \cdot Lat_{max}+(1-(Lat-35)^{2.18}/35^{2.18}) \cdot Lat_{min})+ 0.421 \cdot (Alt^{0.51}/1000^{0.51} \cdot Alt_{max}+(1-Alt^{0.51}/1000^{0.51}) \cdot Alt_{min})+ 0.265 \cdot (Q_{mv}^{0.22}/5000^{0.22} \cdot Q_{max}+(1-Q_{mv}^{0.22}/5000^{0.22}) \cdot Q_{min})$

Month	Q_{max} norm	Q_{min} norm	Lat_{max} norm	Lat_{min} norm	Alt_{max} norm	Alt_{min} norm
1	-0.71	0.58	-1.000	1.040	-0.97	0.47
2	-0.48	0.81	-1.000	1.370	-0.98	0.51
3	-0.17	0.84	-1.000	0.56	-0.58	0.22
4	-0.17	1.580	-1.000	0.38	-0.69	0.24
5	0.62	-0.1	2.170	-0.29	2.110	0.18
6	1.740	-1.000	2.510	-0.23	1.870	-0.32
7	0.52	-1.000	0.63	-0.62	0.51	-0.42
8	0.09	-1.000	0.24	-0.71	0.07	-0.49
9	-0.16	-0.82	0.05	-0.79	0.03	-0.38
10	-0.2	-0.56	-0.03	-0.74	-0.06	-0.2
11	-0.63	0.11	-0.66	-0.28	-0.62	0.07
12	-0.44	0.54	-0.92	0.32	-0.68	0.13

Fig. 8.3. The sub-model for tributary water discharge (compiled from Abrahamsson and Håkanson, 1998).

A smoothing function (see fig. 8.4) is used to average out seasonal variability given by the seasonal variability norm, which is defined to yield extreme values for Q. The equation that specifies this calculation is a smoothing or averaging function, which is based on the five easily accessible factors given above. The smoothing function (SMTH) is contained in the software program "Ithink". The SMTH function uses a first-order exponential equation (fig. 8.4) to smooth the input (here the seasonal variability norm for Q). The SMTH function works in the same way as the one- and two-sided running mean values. It may be written as:

$$\text{SMTH} = \text{SMTH(input, averaging function, initial value)} \qquad (8.1)$$

This function smoothes the seasonal variability norm for Q by applying a specified averaging function, which operates over a specified time interval, to an input (here the seasonal variability norm), given an initial value for that input. The initial value is simply the mean value of the seasonal variability norm for Q (namely 1). Since the initial results depend on this initial value, this choice is not trivial.

In spite of the fact that river discharge is a variable with great temporal and spatial variability, this sub-model has proven to yield good predictions of monthly average Q. The best results of the validation were achieved for rivers with a mean annual discharge in the range 1 - 500 m^3/s. More uncertain predictions are obtained for the smallest and largest rivers. The range is, however, large enough to include most European rivers (88 of 114 in the calibration and 90 of 119 in the validation). So, given that limitation, the model certainly has a wide range of applicability.

Fig. 8.5 exemplifies how differences in altitude and latitude influence the seasonal moderator for Q. The default conditions are given by a lake with a catchment area of 10 km^2, a mean annual precipitation of 650 mm/yr, at an altitude of 75 m.a.s.l. and a latitude of 60 °N. Curve 1 in fig. 8.5 gives the characteristic seasonal variations in Y_Q. If a similar lake is situated at an altitude of 1000 m.a.s.l., it is likely that the precipitation is more evenly distributed over the year, and the seasonal variability in Q is smaller. On the other hand, if the lake is placed at latitude 40°N, there is a more pronounced seasonal variability pattern in Y_Q. Note that this result is based on extensive calibrations and validations based on empirical Q series from many European catchments areas.

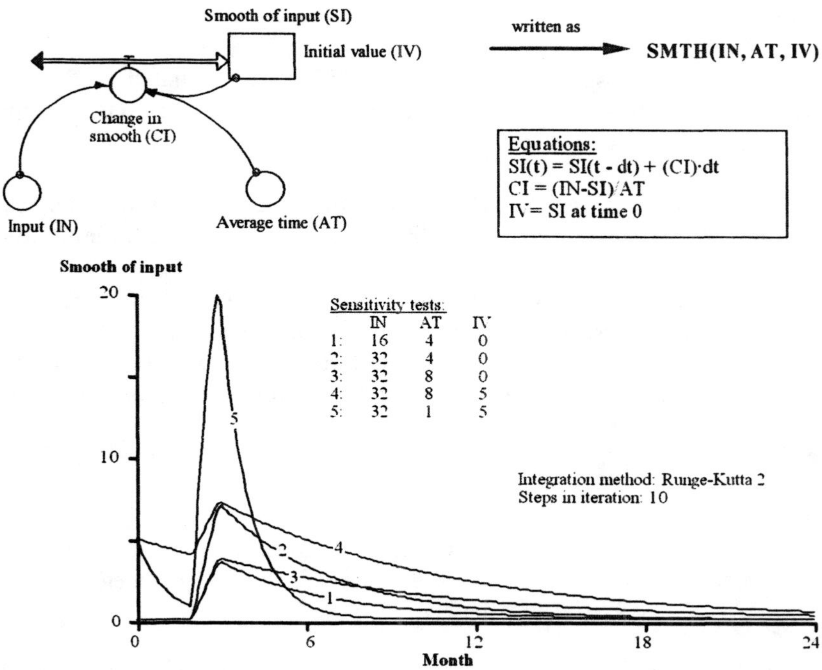

AT = 8

	IV	1	2	3	4	5	6	7	8	9	10	11	12
SI	5	4,41	4,09	7,17	6,33	5,59	4,93	4,35	3,84	3,39	2,99	2,64	2,33
IN	0	0	32	0	0	0	0	0	0	0	0	0	0

Data for curve 4 for 12 time steps

Fig. 8.4. Illustration of the smoothing function.

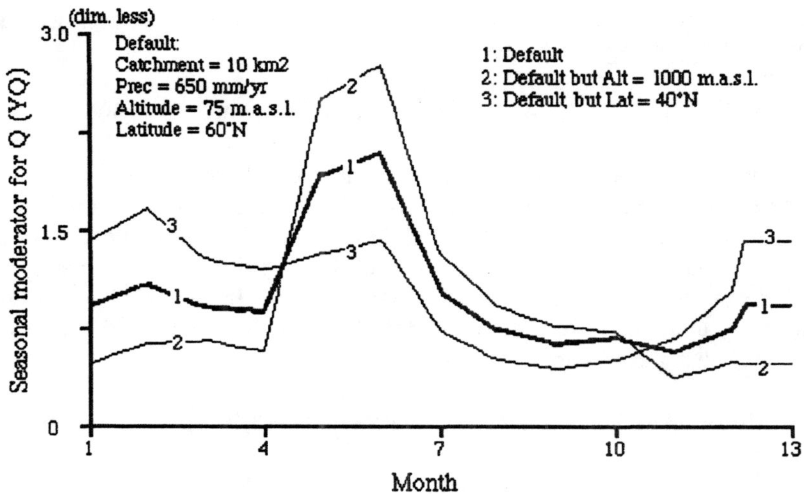

Fig. 8.5. Predicted seasonal patterns in tributary water discharge (Q), as expressed by the dimensionless moderator, Y_Q under default conditions (curve 1), for a catchment at altitude 1000 m.a.s.l. (curve 2) and for a catchment at latitude 40 °N (curve 3).

8.2.2. Lake temperatures

The temperature model (from Ottosson and Abrahamsson, 1998) is illustrated in fig. 8.6. Water (and air) temperatures are governed by many complicated climatological relationships. This approach accounts for the following factors:

- Altitude (Alt, in m above sea level). The higher the altitude, the lower the lake temperature and the greater the seasonal variability in temperature, if everything else is constant.
- Latitude (Lat). The higher the latitude, the lower the temperature and the greater the seasonal variability in temperature, if all else is constant. Latitude is given in °N.
- Continentality, or distance from ocean (in km). The farther away from the Ocean, the more continental, the colder and the greater the seasonal variability in temperature, if all else is constant. The distance is given in km.
- Lake volume (V in m^3). The larger the volume, the smaller the seasonal variability in temperature.

In this approach, both seasonal variability of surface (or epilimnetic) and bottom (or hypolimnetic) water temperatures can be predicted from these four map parameters. This approach uses:

1. A seasonal variability norm for surface water temperature. The monthly data for this norm are given in fig. 8.6. The norm illustrates a standardized case of extreme seasonal variability in surface water temperature. Note that this norm is constructed to be symmetric, but that one obtains realistic values of seasonal variations when the norm is being smoothed.
2. A smoothing function based on the four map parameters is used to level out the norm. A first-order exponential smoothing (SMTH) of the input (the norm for temperature) is used (see fig. 8.5).

The mean annual epilimnetic temperature, MAET (in°C) is utilized as an averaging function. It is calculated from the following expression:

$$MAET = 44 - ((750/(90-Lat))^{0.85})^{1.29} - 0.1\cdot\sqrt{Alt} - 0.25\cdot(Cont^{0.9}+500)^{0.52} \quad (8.2)$$

Where Lat = Latitude (°N); Alt = Altitude (m.a.s.l.); and Cont = distance (in km) from the lake to the ocean. The empirical constants in eq. 8.2 were obtained after calibrations using data for a wide range of lakes (see Ottosson and Abrahamsson, 1998). MAET should be high close to the equator (Lat = 0); the value decreases with the square root of altitude (all else being constant). MAET is rather independent of continental cooling near the oceans [given by $(Cont^{0.9}+500)^{0.52}$].

The mean monthly epilimnetic temperature (MMET = SWT) of a given lake may then be estimated from the following expression:

$$MMET = MAET + SMTH(T_{norm}, (MAET/6)\cdot(V\cdot10^{-6})^{0.1}), 1) \quad (8.3)$$

Where T_{norm} = the norm for surface temperature (see fig. 8.6); V = lake volume (in km^3). The input is T_{norm}. The averaging function is MAET/6. The higher the MAET, the smoother the curve for monthly surface temperature. The factor of 6 is an empirical constant used to get a relevant smoothing, since high values of MAET, like 25 (°C), would produce almost a straight line, and 25/6 = 4.2, gives a more realistic smoothing. The initial value is 1, meaning that one first calculates MAET, then, for the summer period, positive values are added to MAET, and during the winter, negative values are added to MAET. When the lake water is perennially frozen, if MAET < -10, then MMET is set to 0. The size of the lake (V) is also used in the averaging function to smooth the norm (but V will not influence the mean annual water temperature) in such a way that lakes with a large volume will get a smaller monthly variability in surface temperature than small lakes. The lake size variable is given by: $V^{0.1}$. The exponent 0.1 is an empirical constant applied to obtain a realistic smoothing.

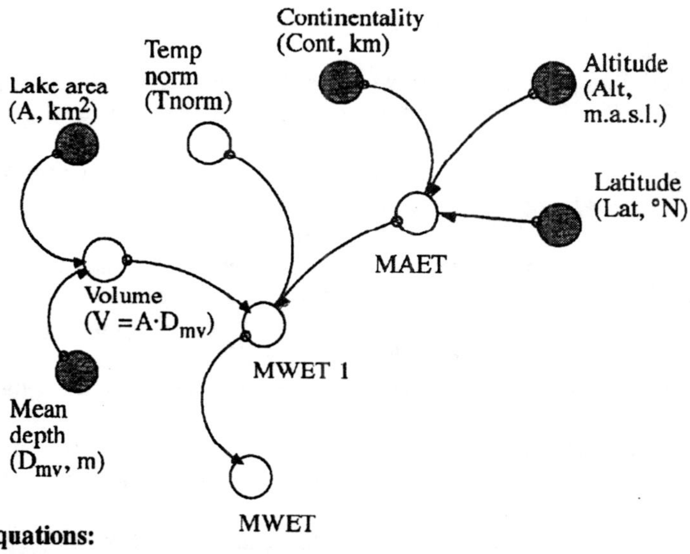

Equations:

MAET = Mean annual epilimnetic temperature, °C

$MAET = 44-(750/(90-Lat^{0.85}))\wedge 1.29-0.1\cdot\sqrt{Alt}-0.25\cdot(Cont^{0.9}+500)\wedge 0.52;$

$MWET = MAET+SMTH(T_{norm},(52/12)\cdot MAET/6\cdot(V\cdot 10^{-6})\wedge 0.1,1)$

If MWET 1 < 0 then 0 else MWET

MWET = Mean weekly epilimnetic temperature, °C

MMHT = Mean monthly hypolimnetic temperature, °C

If MMHT1 < 4 then MMHT = 4 else if MAET < 4 or MAET > 17 or $D_{mv} < 2$ then

MMHT = MAET else

$MMHT = SMTH(MMET1,\sqrt{MAET}/(0.51\cdot MAET)/(0.5/(1.1/(D_{mv}+0.1)+0.2))$

Month:	1	2	3	4	5	6	7	8	9	10	11	12
Temp norm:	-8.0	-2.0	0	2.0	8.0	20.0	8.0	2.0	0	-2.0	-8.0	-20.0

Fig. 8.6. The sub-model for lake temperatures (compiled from Ottosson and Abrahamsson, 1998).

Hypolimnetic temperatures (°C) are characteristic for dimictic lakes, which appear from approximately latitudes 30 °N to 60 °N and altitudes 0 to 4000 m (see Wetzel, 2001). A limit is also set for very shallow lakes at such latitudes and altitudes. Lakes with a mean depth less than 2 m are not generally dimictic but polymictic. For lakes with mean depth greater than 2 m, this approach assumes that the relationship between surface and bottom temperatures depends on the mean depth: The smaller the mean depth, the smaller the difference between surface and bottom temperatures. The curve for the mean monthly hypolimnetic temperature (MMHT = DWT) should be markedly smoother than the curve for the mean monthly surface temperature (MMET). If the predicted mean annual water temperatures (MAET) are higher than 17 (°C), the lake would probably be warm monomictic; if MAET is below about 4 (°C), the lake would be cold monomictic. Between these limits, the hypolimnetic temperatures are predicted from:

$$MMHT = SMTH(MMET, \sqrt{MAET}/(0.51 \cdot MAET)/(0.5/(1.1/(D_{mv}+0.1)+0.2))) \quad (8.4)$$

- The function to be smoothed (the input) is the predicted mean monthly surface temperature (MMET).
- The seasonal variability for the mean monthly hypolimentic temperature (MMHT) should be smaller than for MMET. This influence is given by the averaging value of √MAET.

Fig. 8.7 shows predictions of SWT = MMET and DWT = MMHT with this model, which is meant to be generally applicable for European lakes.

A. Gives the surface and bottom temperatures (SWT and DWT) for a lake at altitude 75 m.a.s.l., latitude 60 °N, continentality 100 km, area 1 km^2 and mean depth 4.3 m. Note the difference between summer and winter values. The lake circulates during spring and fall when these two curves cross, otherwise the lake is stratified.
B. Simulations for a lake at altitude 1000 m (otherwise the same conditions as in fig. 8.7A). Note that the winter stratification is stronger in this case.
C. Gives the same for a lake at latitude 40 °N. Note that the lake is not dimictic (it is warm monomictic, as it should be). It only stratifies during summer, and the temperatures are generally significantly higher, as compared to fig. 8.7A.
D. Simulations for a lake at 1000 km from the ocean. This lake is markedly dimictic.

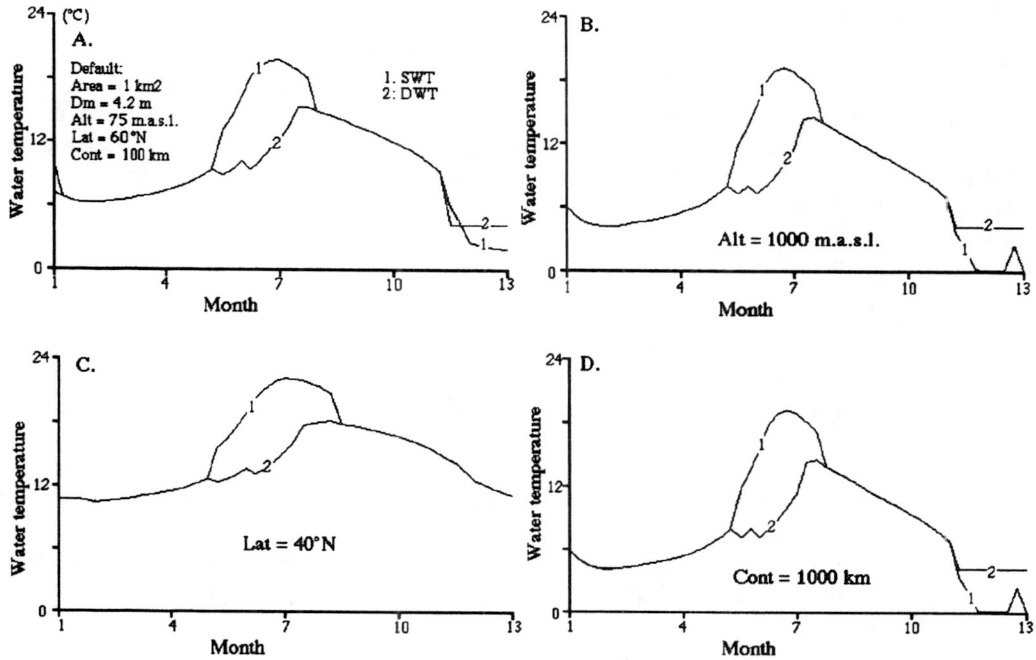

Fig. 8.7. Illustrations of the temperature sub-model. A. Predictions in the dimictic default lake. B. Predictions if this lake is at altitude 1000 m.a.s.l. C. Predictions if this lake is at latitude 40 °N. D. Predictions if this lake is 1000 km from the ocean.

8.3. Information on the 95 lakes

Many of the examples presented in this book are taken from an extensive field investigation involving 95 Swedish lakes studied during 1986-1989 as part of an experimental program to reduce the concentrations of mercury and radiocesium in fish. Data were collected on the concentrations of Hg and ^{137}Cs in pike and perch fry, the load of Hg and ^{137}Cs to the lakes, the different lake properties and the remedial measures. The study lakes were selected to represent as wide a geographical and limnological range as possible, and they were "at risk" from Hg and ^{137}Cs contamination.

- Morphometric data, like volume, area and mean depth were calculated from bathymetric maps. Most lakes are smaller than 1 km^2; the largest lake is 5.26 km^2, the smallest 0.02 km^2. Mean depth ranges from 0.9 m to 10.1 m, with a mean of 4.2 m.
- The following lake variables (discussed in this book) were determined from monthly samples: pH, Secchi depth, temperature, alkalinity, total-P, conductivity, Fe concentration, Ca concentration, hardness (= Ca+Mg concentration) and color.
- Theoretical water retention time and the tributary water discharge (as determined from the specific runoff) are also included.
- Data on the catchment area, for example, the proportion of the catchment area covered by lakes (Lake%), by mires (Mire%), by various types of soils, rocks, and by land uses, etc. were also determined (table 8.1).

Hg determinations were made on individual and pooled samples of muscle from pike and 1+perch (perch in their second year of life), caught in late July, August and September. The pike were caught between January and May. Weight-length relations were used to standardize the data on Hg concentration for size for both pike (1-kg pike; from regressions) and perch (1+perch).

Basic data on the lakes (drainage area characteristics, morphometry and water chemistry) are given in table 8.1, which gives annual mean values for the period March 1986 - February 1987. Table 8.1 therefore describes the conditions in the lakes before the remedial measures were introduced.

The mean annual value for lake pH (pH_{12}) before remediation varied from 4.8 to 6.7; the mean value for all lakes was 5.9. The mean annual lake color (Col_{12}) also varies widely (from 15 to 340 mg Pt/l). Most of the lakes have low alkalinity and a color value around 100, which is typical for glacial boreal lakes. The mean annual lake concentrations of total-P varied between 3.8 and 33.4 µg/l.

Table 8.1. Basic data from the 95 Swedish lakes. SD = standard deviation; n = number of lakes.

	Mean	Min.	Max.	SD	n
A. Drainage area parameters					
ADA (km^2)	15.8	0.36	81	16.6	95
RDA	41.0	8	194	31.9	95
Lake%	2.3	0	13.1	2.8	95
Forest%	80.0	60.9	100	9.1	95
Mire%	14.2	0	35.3	8.5	95
Open land (OL)%	3.4	0	20.8	5.1	95
Rock%	9.1	0	77.0	13.8	95
Till%	67.7	3.2	96.0	19.7	95
B. Lake morphometric parameters					
L_{max} (km)	1.82	0.22	7.80	1.43	
D_{max} (m)	12.4	1.7	52.0	8.9	95
area (km^2)	0.77	0.02	5.26	0.94	95
area/ADA	0.073	0.002	0.391	0.068	95
Vol (km^3)	0.0041	0.00005	0.0495	0.0070	95
B_m (km)	0.35	0.06	1.05	0.20	95
D_{mv} (m)	4.18	0.90	10.10	2.40	95
D_{rel}	1.71	0.18	7.39	1.29	95
L_d	2.43	1.40	8.20	0.96	95
V_d	1.18	0.50	2.20	0.36	95
DR	0.24	0.04	1.13	0.18	95
ET (% of area)	26	7	90	16	95
Q_{mv} (m^3/sec)	0.164	0.010	0.850	0.173	95
T (years)	0.873	0.020	4.790	0.917	95
pH_{12}	5.9	4.8	6.7	0.4	95
alk_{12} (meq/l)	0.048	0.000	0.370	0.056	95
$cond_{12}$ (mS/m)	3.96	1.58	10.88	2.19	95
TP_{12} (µg/l)	11.5	3.8	33.4	5.5	67
Fe_{12} (µg/l)	543	78	1950	358	41
Ca_{12} (meq/l)	0.17	0.08	0.80	0.12	50
$CaMg_{12}$ (meq/l)	0.24	0.09	0.64	0.11	92
Sec_{12} (m)	2.5	1.1	6.3	1.0	88
Col_{12} (mg Pt/l)	98	15	340	53	95

ADA=Drainage area
RDA=Relief drainage area
Lake%=Lake% of ADA (etc)
ET=Bottom areas of erosion & transp.
B_m=Mean breadth
D_{mv}=Mean depth
L_d=Shore development
V_d=Volume development
DR=Dynamic ratio
D_{rel}=Relative depth
Q_{mv} =Mean annual water discharge
T=Water retention time
pH_{12}= Mean annual pH
Sec_{12}= Mean annual Secchi depth
Col_{12}= Mean annual color

8.4 Basic equations in the mass-balance model (LakeMab)

This section gives is the set-up for the abiotic fluxes. Abbreviations: M = mass; F = flow; R = rate; C = concentration; DC = distribution coefficient; T = time or age; Y = dimensionless moderator; D = depth; PF = particulate fraction; ET = ET areas; A = A areas; SW = surface water; DW = deep water; flux from A to B = F_{AB}, etc. Mass in A = M_A, etc. See fig. 8.8 for illustration.

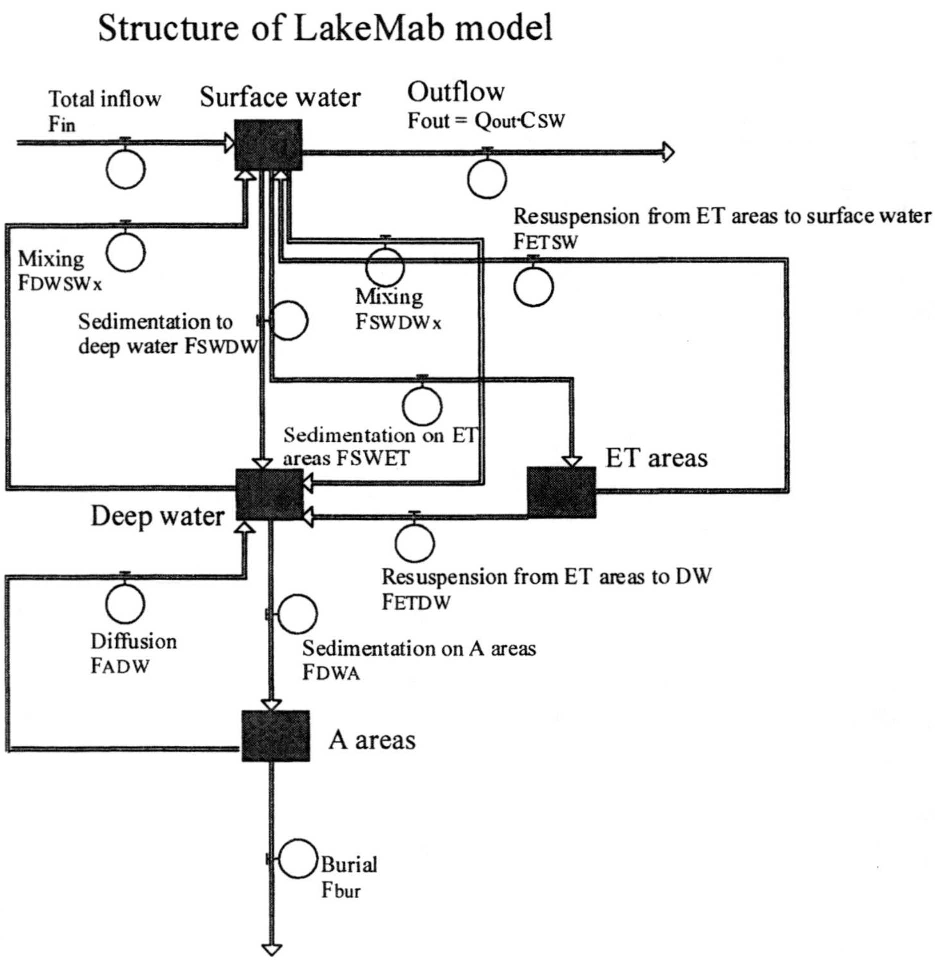

Fig. 8.8. Illustration of the basic features of the abiotic part of the LakeMab model, which includes the following compartments: (1) the surface-water compartment (SW), (2) the deep-water compartment (DW), (3) the ET areas and (4) the A areas.

Compartment, X in surface water (M_{SW}):

$$M_{SW}(t) = M_{SW}(t - dt) + (F_{in} + F_{ETSW} + F_{XDWSW} - F_{out} - F_{SWET} - F_{SWDW} - FM_{SWDWx})\cdot dt \quad (8.5)$$

INFLOWS:

$F_{in} = Y_Q\cdot Q_{mv}\cdot C_{in}\cdot 60\cdot 60\cdot 24\cdot 7$

$F_{ETSW} = (1\text{-}Vd/3)\cdot M_{ET}\cdot 1.386/T_{ET}$

$F_{DWSWx} = M_{DW}\cdot R_{mix}$

OUTFLOWS:

$F_{out} = M_{SW}\cdot R_{out}$

$F_{SWET} = (M_{SW}\cdot v\cdot (1\text{-}DC_{res})\cdot PF\cdot ET/D_{ET})+(M_{SW}\cdot v\cdot 2\cdot DC_{res}\cdot PF\cdot ET/D_{ET})$

$F_{SWDW} = (M_{SW}\cdot v\cdot (1\text{-}DC_{res})\cdot PF\cdot (1\text{-}ET)/D_{ET})+(M_{SW}\cdot v\cdot 2\cdot DC_{res}\cdot PF\cdot (1\text{-}ET)/D_{ET})$

$F_{XSWDW} = M_{SW}\cdot R_{mix}$

Compartment, X in deep water (M_{DW}):

$$M_{DW}(t) = M_{DW}(t - dt) + (F_{SWDW} + F_{XSWDW} + F_{ETDW} + F_{ADW} - F_{DWA} - F_{XDWSW})\cdot dt \quad (8.6)$$

INFLOWS:

$F_{SWDW} = (M_{SW}\cdot v\cdot (1\text{-}DC_{res})\cdot PF\cdot (1\text{-}ET)/D_{ET})+(M_{SW}\cdot v\cdot 2\cdot DC_{res}\cdot PF\cdot (1\text{-}ET)/D_{ET})$

$F_{SWDWx} = M_{SW}\cdot R_{mix}$

$F_{ETDW} = (Vd/3)\cdot M_{ET}\cdot 1.385/T_{ET}$

$F_{ADW} = M_A\cdot R_{diff}$

OUTFLOWS:

$F_{DWA} = (M_{DW}\cdot v\cdot (1\text{-}DC_{res})\cdot PF\cdot (1\text{-}ET)/D_A)+(M_{DW}\cdot v\cdot 2\cdot DC_{res}\cdot PF\cdot (1\text{-}ET)/D_A)$

$F_{DWSWx} = M_{DW}\cdot R_{mix}$

Compartment, X on A areas (M_A):

$$M_A(t) = M_A(t - dt) + (F_{DWA} - F_{bur} - F_{ADW})\cdot dt \quad (8.7)$$

INFLOWS:

$F_{DWA} = (M_{DW}\cdot v\cdot (1\text{-}DC_{res})\cdot PF\cdot (1\text{-}ET)/D_A)+(M_{DW}\cdot v\cdot 2\cdot DC_{res}\cdot PF\cdot (1\text{-}ET)/D_A)$

OUTFLOWS:

$F_{bur} = M_A\cdot 1.386/(T_A\cdot 52)$

$F_{ADW} = M_A\cdot R_{diff}$

Compartment, X on ET areas (M_{ET}):

$$M_{ET}(t) = M_{ET}(t - dt) + (F_{SWET} - F_{ETSW} - F_{ETDW})\cdot dt \quad (8.8)$$

INFLOWS:

$F_{SWET} = (M_{SW}\cdot v\cdot (1\text{-}DC_{res})\cdot PF\cdot ET/D_{ET})+(M_{SW}\cdot v\cdot 2\cdot DC_{res}\cdot PF\cdot ET/D_{ET})$

OUTFLOWS:

$F_{ETSW} = (1\text{-}Vd/3)\cdot M_{ET}\cdot 1.386/T_{ET}$

$F_{ETDW} = (Vd/3)\cdot M_{ET}\cdot 1.385/T_{ET}$

8.5. Basic equations in the foodweb model (LakeWeb)

This section gives an outline of each secondary unit in the LakeWeb model (there are two zooplankton units, zoobenthos and two fish units); and fig. 8.9 gives an illustration of the model and also abbreviations. The basic equation is given by:

$$BM_{SU}(t) = BM_{SU}(t - dt) + (IPR_{SUPU1} + IPR_{SUPU2} - CON_{SUPU1} - CON_{SUPU2} - EL_{SU})\cdot dt \quad (8.9)$$

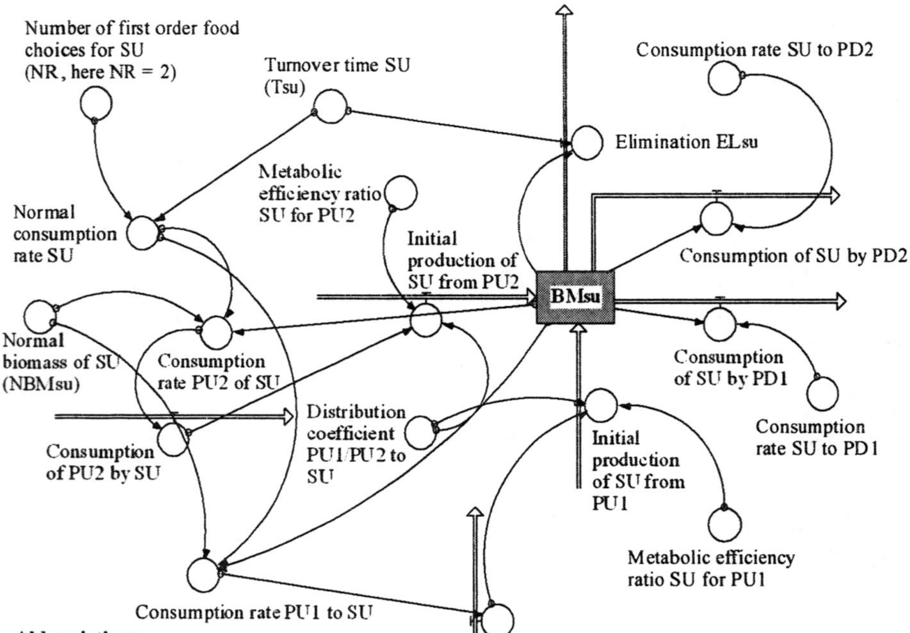

Abbreviations:
BM = Biomass
CON = Consumption (= out)
CR = Actual consumption rate
DC= Distribution coefficients, e.g., for food choices
EL = Elimination related to death
MER = Metabolic efficiency ratio
NCR = Normal consumption rate
NR = Number of first order food choices
IPR = Initial production (= in)
PR = Production
PU1 and PU2 = Primary unit 1 and 2
PD1 and PD2 = Predator unit 1 and 2 (feeding on the secondary unit)
SU = Secondary unit (e.g., herbivorous zooplankton)
T = Turnover time

for fish also:
MIGin = migration into the lake
MIGout = migration from the lake

Set-up using:
• 2 primary production units,
• 2 predatory units and
• 1 secondary unit, the target unit in this set-up

Fig. 8.9. The set-up of each secondary unit in the LakeWeb model. The figure also gives general abbreviations.

Where BM_{SU} is the biomass (BM) of the secondary unit (SU; in kg ww). The initial BM_{SU} value is generally set equal to the normal biomass (NBM_{SU} in kg ww), as calculated from empirical equations in table 4.1.

In the set-up shown in fig. 8.9, the initial secondary production (IPR) is related to two fluxes, from primary units 1 and 2 (PU_1 and PU_2). For each flux, IPR is given by:

$$IPR_{SUPU1} = DC_1 \cdot CON_{PU1SU} \cdot MER_{PU1SU} \quad (8.10)$$

Where DC_1 is the first order distribution coefficient. The LakeWeb model uses a simple general system to assign weights on food choices and adjust the consumption rates for the number of food choices. There are, for example, three food choices for zoobenthos (benthic algae, macrophytes and "sediments"); and prey fish have a menu of three food alternatives predatory zooplankton, herbivorous zooplankton and zoobenthos). In the set-up shown in fig. 8.9, the secondary unit has two first order food choices, so NR = 2. So, if there are more than two food choices, they are first differentiated by a distribution coefficient (DC_1) into two first order food choices and then by a second distribution coefficient (DC_2) into second order food choices, etc.

CON_{PU1SU} is the consumption (outflow or loss) of biomass from the compartment PU_1 from grazing by the animals in compartment SU (kg ww/week). The grazing rate (= the actual consumption rate, CR_{SUPU1}) is calculated from:

1. the normal consumption rate related to the number of first order food choices (NR) and the inverse of the turnover time of the consumer. That is:

$$NCR_{SU} = NR_{SU}/T_{SU} \tag{8.11}$$

2. the normal biomass of the secondary unit (NBM_{SU}), as given by an empirical model (see table 4.1), and
3. the actual biomass of the secondary unit (BM_{SU}), as this number is calculated by the LakeWeb model. That is, the actual consumption rate, CR_{PU1SU}, may be written for each unit as:

$$CR_{PU1SU} = NCR_{SU} + NCR_{SU}\cdot(BM_{SU}/NBM_{SU} - 1) \tag{8.12}$$

Where NCR_{SU} the normal consumption rate for the secondary unit (SU).

The loss of biomass from primary unit 1 from predation by the secondary unit, CON_{PU1SU} is then calculated as:

$$CON_{PU1SU} = BM_{PU1}\cdot CR_{PU1SU} \tag{8.13}$$

The metabolic efficiency ratio (dimensionless), MER_{PU1SU}, gives the fraction of the food that actually increases the biomass of the secondary unit (the consumer). Table 8.2 gives a compilation of characteristic MER values used for all groups of organisms in the LakeWeb model. From this table, one can note that the MER values for plants and animals vary between 0.15 and 0.32. A typical value is 0.2, meaning that 20% of the food consumed increases the biomass of the consumer. The MER value for zoobenthos eating sediments is, however, lower (25% of the MER value for zoobenthos feeding on benthic algae).

Table 8.2. Metabolic efficiency ratios for key functional groups (MER = PR/CON, dimensionless). The MER value is basically calculated from the mass-balance equation, CON = PR + RES + FAE, where CON - consumption, PR - production, RES - respiration, FAE - unassimilated food (faeces) and T – turnover time (= BM/PR, days; BM = biomass). The actual consumption rate, CR, expresses reductions in biomass of prey organisms per unit of time. The values used for PR, RES, FAE and MER are mainly based on data from Winberg (1985). From Håkanson and Boulion (2002)

	CON	PR	RES	FAE	MER	T	CR	Consumes	Consumed by
Zherb	100	24	36	40	0.24	6.0	0.17	Phytopl., bacteriopl.	Zpred, prey fish
Zpred	100	32	48	20	0.32	11.0	0.091	Zherb	Prey fish
Zoobenthos	100	15	35	50	0.15	65	0.015	Macrophytes, benthic algae	Prey fish
Prey fish	100	16	64	20	0.16	300	0.016	Zoobenthos, Zherb, Zpred	Pred. fish
Predatory fish	100	25	55	20	0.25	450	0.0013-0.02	Prey fish	Pred. fish

The initial production of the secondary unit, SU, from consumption of the second primary unit (IPR_{SUPU2}) is handled in the same manner using the same consumption rate but (1-DC_1) instead of DC_1. Then, this initial production is given by:

$$IPR_{SUPU2} = (1-DC_1)\cdot CON_{PU2SU}\cdot MER_{PU2SU} \qquad (8.14)$$

Note that this approach involves several simplifications. One can also use specific consumption rates reflecting species-specific feeding preferences in the LakeWeb model. The loss from the compartment PU_2, CON_{PU2SU}, is calculated in the same manner as the loss from compartment PU_1, CON_{PU1SU}.

The two transport routes from the secondary unit to the two predatory units feeding on the animals in the secondary unit (PD_1 and PD_2) are handled in the same manner. That is, each flow is given as a function of, (1) the normal consumption rates (here NCR_{SUPD1} or NCR_{SUPD2}), (2) the normal biomasses of the two predatory units (NBM_{PD1} and NBM_{PD2}), as these are given by empirical models (see table 4.1), and the actual biomasses of the two predatory units (BM_{PD1} and BM_{PD2}), as these are calculated by the LakeWeb model.

Finally, elimination (EL_{SU} = the loss of biomass in kg ww/week from the secondary compartment) is given as:

$$EL_{SU} = BM_{SU}\cdot 1.386/T_{SU} \qquad (8.15)$$

where 1.386 is the halflife constant. Table 8.2 also gives a compilation of turnover times used for all nine groups or organisms included in the LakeWeb model.

The LakeWeb model also accounts for the fact that fish migrate between habitats for feeding and spawning (Harden-Jones, 1968; Northcote, 1978; McDowall, 1988; Wotton, 1990; Brittain and Brabrand, 2001). One should expect a net inflow of fish via up- and down-stream rivers if the actual fish biomass (BM_F) is lower than the normal fish biomass (NBM_F), since fish are likely to migrate to and stay in better habitats, and vice versa (Busch and Sly, 1992). The migration is also a function of the tributary water discharge (Q_{mv}) and hence also of the theoretical lake water retention time ($T = V/Q_{mv}$).

Macrophytes generally appear with relatively high biomasses in lakes, and they can play several important roles in lake ecosystems. These roles include binding nutrients (handled in the LakeWeb model by the mass-balance model for total phosphorus), serving as a substrate for zoobenthos (handled in LakeWeb in the sub-model for zoobenthos), and providing shelter for fish and thereby increasing fish production and biomasses. It is important to note that the latter is not related to fish feeding on macrophytes (although single species, like carp, do), but indirectly, by providing more protected living conditions for the small fish. This process is also included in the LakeWeb model.

9. Index

Printed in the United States
93327LV00002B/8/A